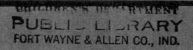

Date Due

DEC 14 1954		
MAR 8 1955		
MAY 3 1955		
MAR 2 0 1956		
MAR 2 7 1956		
MAY 1 5 1956		
MAY 2 2 1956		
13 Feb '57		
MAR 2 6 1957		

INDIANA
BIRDS

This publication was designed with the purpose of aiding more persons in becoming acquainted with the habits and characteristics of common Indiana birds. It is not intended to be a technical booklet for the scientist, but rather a booklet to which the amateur bird-watcher may refer for identification of birds not immediately recognized.

If just one additional Hoosier learns to identify ten birds through the use of this publication, we will feel justified in publishing it.

KENNETH R. COUGILL, *Director*
Division of State Parks, Lands & Waters.

INDIANA BIRDS

By

ALDEN H. HADLEY

Issued by

THE INDIANA DEPARTMENT OF CONSERVATION

Division of State Parks, Lands and Waters

June, 1948

Price Ten Cents

INTRODUCTION

The forty-eight birds described in this publication are considered permanent resident birds of the state. That is they are found in some part of Indiana at all seasons of the year.

Descriptions of the birds were prepared by Alden H. Hadley, nationally recognized ornithologist and former representative of the National Audubon Society. Now employed as a lecturer by the Public Relations Division of the Indiana Department of Conservation, Mr. Hadley has a wealth of experience to draw upon for these articles since he has been in the field nearly a quarter of a century.

The bird illustrations appearing in this publication were taken from the late Major Allan Brooks' true-life color paintings, and are used by special permission of the National Audubon Society. Four of the cuts, however, appearing on pages 34, 36, 44 and 48, were obtained from other sources.

Herbert Sweet of Carmel, Indiana prepared the sketches which appear on pages 53, 54 and 55.

★ ★ ★ ★

TABLE OF CONTENTS

EASTERN BLUEBIRD
(Sialia sialis sialis)

No doubt most ardent lovers of birds could readily name one or more of our native song birds which are held in just a little higher esteem than others. The bluebird has for many years held first place in my interest and affection among all our native song birds with which I am acquainted. Others with whom I have shared my predilection have agreed with me.

The bluebird, which averages 7 inches long, is a blithe and joyous member of the thrush family and it is quite doubtful if there is any other of our commoner native song birds that has so completely escaped being accused of objectionable habits and has thus escaped persecution. With "the sky upon his back and the earth upon his breast" the bluebird returns each spring to old familiar haunts. Maurice Thompson, our famous Hoosier poet and novelist, has given us a true and charming picture of our bird in his poem "The Bluebird", the first stanza of which is as follows:

"When ice is thawed and snow is gone,
And racy sweetness floods the trees;
And snow-birds from the hedge have flown,
And on the hive-porch swarm the bees,
Drifting down the first warm wind
That thrills the earliest days of Spring,
The bluebird seeks our maple groves
And charms them into tasseling."

Often on still and sunny days in late February and early March, the bluebirds' mellow quavering notes may be heard far up in the sky as they journey northward hard on winter's trail. They know the blustery days of March and weather many a belated snow-storm. However, once in a long while they are overtaken by a real tragedy as during the winter and spring of 1894-95. Unusually large numbers of bluebirds remained in the north until late December as the weather had been unusually mild. Then they were forced southward by a severe cold spell.

Then it turned warm and the bluebirds as well as other migrants started northward. Again they were caught in a record-breaking snow and sleet storm which reached almost to the Gulf Coast. For many years the bluebird population was at an extremely low ebb and it is doubtful whether this well-loved bird has even yet recovered anything like its former numbers. The introduction of the English starling, perhaps, has been another factor in causing a decrease in the bluebird population on account of the aggressive habit of the alien species and its tendency to occupy nesting boxes and natural cavities in trees.

The food of the bluebird during the summer months consists of a great variety of insects such as beetles, flies, caterpillars, spiders, ants, and tree-hoppers. The vegetable food consists chiefly of wild fruits. The range is throughout Eastern North America south to southern Florida. The nest is of grasses in a tree cavity or in a box. The eggs are 4 to 6, plain white or bluish white.

NORTHERN BLUE JAY
(Cyanocitta cristata cristata)

The blue jay in the estimation of many bird students, and other persons as well, is a sort of Dr. Jekyl and Mr. Hyde of the bird world. In other words he is a perpetual study in dual personality. Sometimes he seems to be Jekyl and at other times Hyde. He is, for example, held in ill esteem when he comes as a marauder to the nests of other birds and perhaps destroys the eggs and the young as no doubt he sometimes does. However, one naturalist has observed that every living creature is predatory and that man himself, especially so called civilized man, is the greatest predator of all. So perhaps we shouldn't become too disturbed or excited when a jay despoils a robin's nest or that of some other bird in our dooryard, for certain other highly regarded birds manifest the same trait.

However, we all will have to agree that the jay is a handsome fellow with his blue jacket and the black collar about his neck. And when he goes trooping through the autumn woods in company with titmice and chickadees he certainly is calculated to arouse the interest and admiration of most bird-lovers.

Perhaps the casual bird student is not aware that the jay, to some extent, is both a mimic and a ventriloquist. In addition to his rather harsh and strident well-known notes he now and then indulges in a genuine musical performance. This consists of a series of subdued, scarcely audible notes which seem quite out of keeping with his family relationship for he is first cousin to the crow.

Edward Howe Forbush in his "Birds of Massachusetts" has given a vivid and impressionistic picture of the blue jay. "The handsome, active blue jay is an engaging rascal. Where there are blue jays there is always action and usually noise, for jays like crows are fond of hearing their own voices. Often a great uproar in the woods may be traced to a dozen or more blue jays in the tree tops, screaming as if in great terror or pain, and apparently for no earthly reason except to keep up the excitement which seems to be of the hair-raising kind, and to exercise their lungs; in autumn they apparently delight in gathering near some woodland dwelling and yamping in a raucous chorus apparently with no object in view except to wake the sleepers; but let them find a screech owl dozing the day away close to the trunk of some tall pine—then we see real excitement! The woods ring with the screaming chorus and blue flashes to blue as the crested birds converge to the attack."

The blue jay is from eleven to twelve inches in length. It ranges from Canada to the Gulf States. Two sub-species are recognized by the American Ornithologists' Union, the Florida blue jay and Semple's blue jay. Aside from any objectionable habits the jay is known to be of considerable economic importance as a destroyer of harmful insects.

The nest is of small twigs lined with rootlets. The eggs are four to six, pale olive green, speckled with brown.

EASTERN CARDINAL
(Richmondena cardinalis cardinalis)

The cardinal is perhaps the most beautiful and conspicuous member of the finch family in the United States. Six more or less varying forms have been described and recognized by the American Ornithologists' Union. These occur all the way from New England to Florida and to Texas, Arizona and New Mexico. However, the average student of birds would doubtless note no differences between these local races and would be quite content just to admire this fine finch and to rank him high among the list of our Indiana song birds. Our Indiana bird is known as eastern cardinal and ranges from South Dakota, southern Iowa, northern Indiana, northern Ohio, south-western Pennsylvania, south to the northern parts of the Gulf States.

Our own state, as is well known, has taken appropriate action and designated this familiar songster as our state bird. It is quite generally distributed over the southern half of the state and during the past thirty or forty years seems to have been gradually extending its northward range. It is resident wherever found and is only migratory to the extent that it often has the habit, during the winter of congregating in considerable numbers in dense tangles of wild roses, brambles and low thickets.

This may well represent a local migratory movement of a very few miles. I shall never forget an experience one day in December some years ago. My young son and I were making our annual Christmas bird census. On a farm next to ours was a four or five acre thicket like the one described above. Here we were surprised to find the largest wintering number of cardinals we had ever seen. It was of course difficult to make an accurate count under the conditions but we were reasonably certain that the number was between fifty and sixty. Doubtless they had congregated for protection from winged predators and also to feed upon the fruit of both the haws and the roses.

The song of the cardinal is known and recognized by almost everyone, even the most casual observer. The clear ringing notes may be heard from late February and early March until well towards the end of August. I shall never forget an instance of a cardinal singing in late winter. It was the last week of February, there had been a succession of bright sunny days during which the temperature had risen abnormally. This was followed by a sudden drop accompanied by a snow storm. The air was full of large swirling flakes and as I tramped across the fields to the haw thicket I was greeted by the loud, mellow whistle of a cardinal. Undaunted by this sudden turn in the weather he was telling all the world to be of *"Good Cheer, Good Cheer."* The cardinal will average 8½ inches in length. The nest is of twigs, strips of bark rootlets and dead leaves. From 3 to 4 bluish-white eggs, specked with brown are laid. Although primarily a seed eater, the cardinal consumes many harmful insects.

CEDAR WAXWING
(Bombycilla cedroum)

Most persons who are acquainted with our native song birds will agree that the cedar waxwing is, in a special sense, a bird of unusual beauty and charm. It is from 6½ to 7½ inches long and not only possesses a graceful form combined with delicate shades of silky brown for the major portion of its coat, but it also has certain interesting habits which still further render it a bird of real distinction. Since waxwings are gregarious only during the nesting season and they wander widely about, you never know when or where you will find them. In fact they are real vagabonds among the birds. I once counted 250 waxwings perched on the telephone wires in an East Coast Florida town. That is a record number in my experience with this bird. The flocks usually number from 5 or 6 up to 50 or 60. One April morning sometime ago, from my hotel window in Vincennes, I watched a group of 15 waxwings as they were making a vain attempt to drink and bathe at a bird bath in the lawn. There had been a belated cold spell and a thin layer of ice covered the bird bath and after repeated efforts to solve the problem which confronted them they finally gave up the attempt. During the performance it was interesting to watch the graceful birds as they raised and lowered their conspicuous crests by way of expressing their emotions, one fancied. Their beautiful satiny coats with the red, wax-like tips on the wings still further added to the charm of the picture.

The graceful birds fly in close formation as they go wheeling about from one feeding ground to another. Our ornithologists have vied with one another in attempts to describe the beauty, charm and interesting habits of the waxwing. One observer watched a group of 5 or 6 of these birds perched on a limb close together while a cherry was passed along the line from the beak of one bird to another. W. L. Dawson in his "Birds of Ohio", has written, "It is as though you had come upon a company of Immortals, high-removed, conversing of matters too recondite for human ken, and who survey you with Olympian disdain". The famous and gifted ornithologist, Dr. Elliot Coues, once wrote of the waxwing, "They lead their idle, uneventful lives— debonnaire birds, sociable but not domestic, even a trifle dissipated, good-natured enough to a friend in a scrape, very reliable diners out, and fond of showing off their dressy top-knots on which so much of their mind is fixed."

In addition to their love for cherries and also for various wild fruits they eat great numbers of harmful insects such as caterpillars and inch-worms. They range throughout North America and breed throughout most of their range.

The nest is in trees and is a compact structure of twigs, mosses, strings and wool. The eggs are 3 to 5 bluish-gray or stone color, speckled with black and dark brown.

BLACK-CAPPED CHICKADEE
(Penthestes atricapillus atricapillus)

Probably few birds are held in higher esteem by the ever increasing number of bird-watchers than the various members of the chickadee family. In Indiana we have both the black-capped and the Carolina chickadee. Since these birds are somewhat alike in appearance and their habits are identical they here will be treated together save for a statement with reference to the slight differences between the two birds.

The black-capped chickadee is from 4¾ to 5¾ inches in length and has the greater wing-coverts, edged with whitish, whereas the Carolina chickadee lacks the whitish edging of the greater wing-coverts, and the tail is shorter than the wing. It is from 4¼ to 4¾ inches in length. It is quite doubtful whether the average person who makes a practice of attracting birds to the feeding station in winter can distinguish between these two close relatives. The black-capped chickadee, shown below, is a common resident in the northern part of the state and also is a fairly common winter resident some distance southward. However, it is seldom found in the southern part of the state. The Carolina chickadee is a common resident throughout southern Indiana where it replaces its close relative, the black-capped.

Chickadees are hardy and resourceful birds and are widely distributed, in their varying forms, over most of the United States and Canada. The American Ornithologists' Union has recognized 21 forms which belong to the genus *Penthestes*.

When warblers and thrushes and the "sun-loving" swallows have left us and gone to their winter homes in warmer climes the chickadee comes into his kingdom and his saucy notes ring out in the bleak winter woods and bring us a message of optimism and good cheer. He also comes confidingly about our homes, performing his truly amazing acrobatic feats in the trees of our dooryard, enlivening the winter day with his presence and bidding us be of good cheer. He comes to the lump of suet in the feeder on the limb of the apple tree, often accompanied by his mate, and quickly fortifies himself to withstand the zero morning by dining on the food that has been provided.

It must not be assumed, although the chickadee is one of the first and most frequent visitors to our feeding stations, that he is not abundantly able to take care of himself even under stress of severe winter weather. Among the injurious insects eaten in winter are bark-beetles, curculios, the larvae of coddling moths, the eggs of canker worms and tent caterpillars, as well as those of many kinds of plant lice. It is doubtful if any one of our native birds has a record that exceeds that of the chickadee from the point of view of its economic value.

The nest is in holes in trees, old stumps and in bird houses. It is lined with moss, grass, feathers and other soft materials. The eggs are from 5 to 9, speckled chiefly at the larger end with brown.

EASTERN CROW
(Corvus brachyrhynchos brachyrhynchos)

The crow is, perhaps, better known throughout the length and breadth of the land than any other bird. Like the English sparrow and the starling he is very aggressive and resourceful and adapts his way of life to almost any conditions. Although this black marauder has few friends, his sharp wits and cunning have made it possible for him to survive all the "crow shoots" and other control measures that have been directed against him, even the occasional charges of dynamite that have been touched off in his extensive winter roosts.

Among the many accusations that have been made against the crow it is well that a few be enumerated. Often when field corn is in the "roasting ear" stage crows have proved very destructive. They also are known to pull up the tender sprouting corn. I have often observed them doing this and I have known of farmers arming themselves with shot guns in order to drive away the birds. In former years, before poultry raisers had developed the present artificial methods of propagation, crows were an ever-present menace on farms where they robbed many a nest and occasionally carried off young chicks.

In the lawn of our Morgan county farm there stands a clump of tall pines. Early each spring the bronzed grackles returned to nest in this favorite spot. The crafty and all-seeing crows then began to make regular and frequent visits to the pines in order to take both the eggs and young. I had early learned, as a farm lad, that a dead crow swinging by a foot from the top of a tall pole served well as a warning to other crows who with evil intent came to the pines, and also to the poultry yard. So each year any visitor to our farm home would usually see one or more crows swinging from this improvised gibbet. Crows have been reported to be extremely destructive on the vast breeding grounds of our wild waterfowl in the Canadian provinces of the Northwest. Here they resort in great numbers and eat the eggs of the nesting ducks. In Florida the southern crow and fish crow have proved at times very destructive to the eggs of various species of the heron family.

Many instances could be cited giving evidence of the damage done by crows, even to the extent of killing as many as 200 young lambs on Martha's Vineyard Island, as reported by the eminent ornithologist, Edward Howe Forbush.

Now what can be said in favor of the crow, if anything at all? It is known to eat large numbers of grasshoppers, cut worms, also spiders, many beetles, grub worms, and many other insects. Our eastern crow, which will average about 19 inches in length, ranges from Mackenzie, Manitoba and Newfoundland south to the northern portion of the Gulf States. Three sub-species are recognized. The nest is of sticks lined with grass and strips of bark, well up in trees. The eggs are 4 to 6, bluish-white marked with shades of brown.

9

EASTERN MOURNING DOVE
(*Zenaidura macroura carolinensis*)

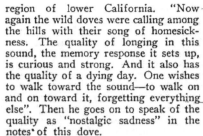

The mourning dove, also called Carolina dove and turtle dove, unlike its now extinct relative the passenger pigeon, appears to be one of nature's great successes from the viewpoint of its survival capabilities. Unlike its famous near of kin, which nested in great numbers in comparatively restricted areas, the mourning dove usually nests solitary although occasionally a few pairs may be found nesting in close proximity.

From time immemorial the dove has won its way into the high regard and affections of mankind. It has figured both in sacred and in our so-called profane literature as a symbol of innocence and peace. Robert Browning and our own Sydney Lanier have given us charming, poetic verses in praise of the dove. And quite recently John Steinbeck in that delightful book, "Sea of Cortez," has recorded his emotional reaction as he listened to the doves (presumably ground doves) as he heard them calling in the lonely semi-desert

region of lower California. "Now again the wild doves were calling among the hills with their song of homesickness. The quality of longing in this sound, the memory response it sets up, is curious and strong. And it also has the quality of a dying day. One wishes to walk toward the sound—to walk on and on toward it, forgetting everything else". Then he goes on to speak of the quality as "nostalgic sadness" in the notes' of this dove.

Although our eastern mourning dove, especially in the more northerly portion of its range, has come to be highly regarded by large numbers of people just from the viewpoint of its sentimental and esthetic value, it has been placed in the list of migratory game birds under the Migratory Bird Treaty Act. However, each state through discretionary power, has the liberty of going beyond the Federal regulations in the matter of further restrictive action. So Indiana, exercising this prerogative, has never permitted an open season on doves. It is quite likely that a storm of protest would be forthcoming should an attempt ever be made to grant such an open season. In the South where the shooting of doves has been a popular sport, many abuses have arisen and Federal officials have experienced much difficulty in combatting the practice of shooting over "baited fields".

The mourning dove, which is from 11 to 13 inches long, is a common summer resident throughout Indiana and is generally a fairly common winter resident over most of the southern counties.

Often under stress of exceptionally severe winters they experience much hardship, even to the extent of having their toes frozen off.

The eastern mourning dove breeds from southern Canada and southern Maine south to the Gulf Coast and west to Kansas and Iowa. The nest is of twigs in a bush or tree, or stump, often on the ground. The eggs are 1 or 2 and white.

NORTHERN FLICKER
(Colaptes auratus luteus)

The flicker is one of that small list of common birds that is known and recognized by almost everyone, especially by those who live in the country or in villages. It is perhaps best known throughout its range as yellow hammer, although it is also called golden-winged woodpecker, high hole, wickup and many other names too numerous to mention.

Next to the red-belly and the redhead, it is one of the most strikingly attired of any of the members of the woodpecker family. The white rump and upper tail coverts are a conspicuous field mark, while the broad black crescent on the breast gives a touch of added distinction to a bird which even without these markings is calculated to attract the attention of even the most casual observer. Then when to the male there is added the jet black streaks, which like a moustache run backwards from the base of the bill, it will be agreed by most bird students that the flicker is really a Beau Brummel.

Not only is the flicker a bird of unusually striking appearance but a small volume could well be written concerning its interesting habits and its life history. It averages about 12 inches in length and unlike its relatives it, to a great extent, has become a ground feeding bird which perhaps renders it more easy to observe, for it comes and confidingly hops around the lawn with the robins and other birds of our dooryards. It will go to an ant hill, fill its crop almost to bursting and if there be young in a nearby nest it then feeds them.

Prof. Beal of the U. S. Department of Agriculture, who years ago made careful studies of the food habits of birds, has recorded in one of his reports that 5,000 small ants were found in the stomach of one bird and 2,000 in each of two others. The flicker also eats numbers of beetles, grasshoppers, caterpillars, spiders, myriapods and many other insects most of which are harmful. Wild fruits are eaten, especially in late summer and early autumn, such as wild cherries and sour gum. The fruit of the poison ivy and the bayberry are favorite winter foods.

Among those who have investigated the food habits of our birds the flicker has been given high rank as a bird of considerable value. William Dutcher, first president of the National Association of Audubon Societies, once wrote as follows: "If the flicker had no other valuable economic quality it would deserve protection because it is the enemy of the ant family, fifty per cent of its food for the year being of these insect pests."

The flicker has the greatest variety of notes and calls of any member of its family. The most common is perhaps *wake up, wake up*. The courtship is also a performance of very great interest.

The northern flicker ranges throughout eastern North America from Alaska and Laborador south to the Gulf States. The nest is usually in a hole in a tree or stub. The eggs are 5 to 9 and glossy white.

11

EASTERN GOLDFINCH
(Spinus tristis tristis)

It is difficult to speak with due restraint when one starts to write about this charming little member of the finch family which although officially known as goldfinch, is also commonly called lettuce bird, thistle bird, yellow bird, and wild canary. Someone has said that this little finch is "panoplied in jet and gold", and this description as applied to the male in breeding plumage has both poetic charm and scientific accuracy. My first acquaintance with this bird was made on the farm when, as a lad, I learned to know them as they came in numbers to pilfer the seeds of sunflower and lettuce from the late summer gardens. Again they flocked about the thistles in the wood-pasture, not only to use the seeds for food but also for the purpose of nest-building.

The food of the goldfinch consists not only of the seeds mentioned, but also of a great variety of plants such as dandelion, goldenrod, aster, wild clematis, cosmos, burdock and catnip. Sometimes

in mid-winter a small flock may be seen swinging like miniature acrobats on the pendent balls of the sycamore on which they are feeding.

One can think of no more charming outdoor picture than this on a cheerless winter day when the fields and woods are snow clad and a biting wind is blowing. One would not be inclined to believe that such a dainty, graceful and resourceful bird would sometimes meet a tragic fate while dining upon one of its favorite foods. Nevertheless such is now and then the case. I allude to the fact that instances are on record of the bird being caught and helplessly imprisoned amidst the clinging hooks of a burdock head. One instance of this came under my observation. Its fondness for the seeds of the sunflower sometimes gets the bird into disrepute upon the part of those who raise these plants for commercial or other purposes.

Thus far we have spoken of the goldfinch as a seed-eater. In the spring it consumes many insects such as young grasshoppers, beetles, inch-worms and plant lice and their eggs.

The home-life of the goldfinch has about it much of interest and charm. The nest building is delayed much later than in the case of most other birds for it usually takes place in the months of July and August. It has been suggested, inasmuch as its favorite nesting materials consist of plant fibres and other soft materials such as thistle down, that this postponement to late summer is due to the birds instinctive preference for such things. According to Edward Howe Forbush "the nest is built chiefly or wholly by the female while the male accompanies her in her labor, caresses her and cheers her with song." Six forms of the goldfinch are recognized by the American Ornithologists' Union. The eastern goldfinch which averages 5 inches in length, ranges throughout Eastern North America and is resident in Indiana, although rare some winters northward. The eggs are 3 to 6, pale bluish-white.

BRONZED GRACKLE
(*Quiscalus quiscalus aeneus*)

Perhaps some will question the inclusion of the bronzed grackle in the list of permanent resident birds of Indiana. However, a few often remain, especially during mild winters, and range over the counties of the southern portion of the state. Years ago E. J. Everman reported them as wintering in Monroe County and they were noted as wintering at Bicknell by E. J. Chansler.

The bronzed grackle, which is also known as crow blackbird and common blackbird, is about 13 inches in length and is a conspicuous member of the family of which he is a member. In his iridescent coat of "polished bronze and blued steel" the male is really gorgeously attired and as he struts about your lawn, taking short quick steps with his head well up, he is a creature calculated to attract attention, were he and his sable clan not so numerous. Grackles seem to have a special preference for pines in which to build their nests and there are few farm homes within the state which have clumps of pines where the grackles do not return each spring, at the appointed time, and begin their nesting activities. Here their raucous music may be heard from early April until June by which time their nesting period is usually over. Most ornithologists agree that the notes uttered by the grackle are rather harsh and somewhat disagreeable. However, this may be, I have always found a certain element of entertainment in their rather harsh and throaty notes.

Dr. Charles W. Townsend has described the courtship of the grackle in part as follows:

"The courtship of the bronzed grackle is not inspiring. The male puffs out his feathers to twice his natural size, partly opens his wings, spreads his tail and, if he is on the ground, drags it rigidly as he walks. At the same time he sings his song—such as it is—with great vigor and abandon."

Grackles are omnivorous in their feeding habits. About one-third of their food has been estimated to be animal and the other two-thirds vegetable. Among animal food are spiders, myriapods, sowbugs, snails, small frogs, tadpoles, small snakes and crawfish. They also consume enormous numbers of insect pests, such as May beetles, grasshoppers, locusts, caterpillars, brown-tail and gypsy moths, cutworms and army worms. They are especially fond of the big fat grubworms that the plowman exposes as he turns his furrow in the spring. I have often observed them as they followed a plowman round after round not a worm escaping their keen vision. They also are known to eat the eggs and young of other birds.

The bronzed grackle ranges throughout eastern North America from New England south to the Gulf States and west to the Rocky Mountains. The nest is usually in coniferous trees, bulky, of grass and mud. From 4 to 7 eggs are laid. These are bluish, or greenish white, blotched and streaked with brown.

13

EASTERN RUFFED GROUSE
(Bonasa umbellus umbellus)

I shall never forget my first meeting with the ruffed grouse. As a farm lad in the early teens I was slowly and noiselessly stealing through my favorite woodland in the hill country of Morgan County in search of squirrels. It was a virgin forest and only in places did the sun's rays penetrate the green canopy overhead. All of a sudden I was startled by a shadowy bird-form which sprang from the cover at my very feet and with a breath-taking whirr of wings disappeared in the half-gloom of the forest. Years afterwards I became more intimately acquainted with the ruffed grouse in other parts of the country, but this memory has lasted all through the years.

This fine game-bird, which is also sometimes called partridge and pheasant, is from 15½ to 19 inches in length with a wing-spread of from 22 to 25 inches. The male is really strikingly attired with his crested crown and partially concealed ruff and his banded tail. The ruffed grouse is a bird of the forest and is quite out of place and ceases to thrive in other habitats. Ornithologists have described with enthusiasm, and in detail, the drumming performance of the male which occurs for the most part during the mating season. This usually takes place from the vantage point of a log or stump and is produced by the rapid vibration of the wings against the feathers of the breast. The male also often struts like a turkey cock as a display performance.

Now we may ask about the past, present, and probable future status of this fine game bird. Butler in his "Birds of Indiana" (1897) reports it as "occurring in varying numbers throughout the state" and gives statements from numbers of observers which indicate that its principal haunts were the heavily wooded hill country of the southern portion of the state. Now what of the present status of this bird in Indiana and what prediction may be made for the future? We are indebted to William B. Barnes of the Pittman Robertson Project for the most recent report on the ruffed grouse within the state. Those interested should read his valuable article in the Indiana Audubon Year Book for 1944. He states, "only in the southern part of the Crawford Upland has this fine bird continued to persist" and his studies show that since the time Butler wrote, this grouse has shown a steady decline in numbers and its future is uncertain.

Dr. Arthur A. Allen of Cornell University for many years conducted experiments in the artificial propagation of this fine game bird. These experiments attracted the attention of wildlife conservationists throughout the country, but his findings revealed that artificial propagation of this game bird is impractical.

In its six recognized forms it ranges across the continent from Alaska and the Canadian Provinces south to the northern tier of the Southern States. The nest is in a slight depression at the base of a tree or stump. The eggs are 8 to 12, and pinkish buff.

BROAD-WINGED HAWK .
(Buteo platypterus platypterus)

This interesting and generally useful member of the hawk family is probably less known to the average student of bird-life than most species of the group to which it belongs. In body conformation it closely resembles its near of kin the red-tailed hawk. The male measures from 13¼ to 16½ inches in length with a wing-spread of from 32 to 38 inches, while the female is from 15 to 18¾ inches long with a wing-spread of from 33½ to 39 inches. Aside from this difference in size the sexes look alike.

Dr. John B. May, in his well-known and authoritative work, "The Hawks of North America," published by the National Audubon Society, writes in part as follows of the broad-wing: "The broad-wing is one of the tamest and most unsuspicious of hawks, often allowing quite close approach. It spends much of its time sitting quietly among the branches of a tree within the forest, sometimes with hunched shoulders and drooping head, and it is doubtless often passed by unobserved when its more active relatives would be noted. It is a typical small Buteo, showing all the characteristics of the group except that it soars less at a considerable elevation or for a prolonged period. It has the chunky body, short square-ended or "fan-shaped" tail, fairly long but broad and round-ended wings, and sluggish movements of its tribe. When hunting, it sometimes hovers like a sparrow hawk or a diminutive rough-leg. Its ordinary call, given both when in flight and when at rest, is a drawling, slurred, high-pitched whistle *chuck-ke-e-e-e* or *psswhee-e*, much like the note of a wood peewee; it has been likened to a steam whistle that "peters out" at the end."

A number of ornithologists have, from time to time, recorded interesting and valuable observations concerning this bird's habits and behavior. Most of these accounts emphasize the comparative tameness of this hawk. Forbush relates that Audubon's "brother-in-law climbed to a nest and took in his hands nest, eggs, bird and all, merely cover-

ing the bird with his handkerchief. Audubon asserts that he carried the bird home and that it sat quietly on a perch while he measured and drew it." C. J. Maynard tells of a broad-wing of which he went in pursuit, and says "that he walked to within a few yards of it while it sat on a low limb." No doubt these are unusual instances which serve to portray the generally tame and inoffensive nature of this hawk, for after all it is a true hawk and is a successful hunter of the prey which constitutes its food. This consists mostly of mice and other small mammals, reptiles, bratrachians, crawfish, earthworms and many kinds of insects.

The broad-wing ranges widely over eastern North America and south to northern South America. It is resident in southern Indiana. The nest is in trees, of sticks, lined with leaves and strips of bark. The eggs are 2 to 4, grayish-white, blotched with various shades of brown.

15

COOPER'S HAWK
(Accipiter cooperi)

Among birds of prey none is a more stealthy and merciless killer than the cooper's hawk. It is a veritable whirlwind of dauntless ferocity. It, also, to a great extent, is a forest inhabiting bird. Forbush, in his Birds of Massachusetts, says of this hawk: "It is cradled in the wind-swept woods and fledged amid the creaking and groaning of great trees. Alert, swift and dauntless it roams the green wood with falcon-like freedom, carrying terror to the hearts of weaker creatures and leaving behind it a trail of destruction and death."

The cooper's hawk, in most respects, closely resembles its smaller relative the sharp-shinned hawk. The male is from 14 to 18 inches in length with a wingspread of from 27 to 30 inches, while the female is usually much larger than the male, being from 16 to 20 inches in length, with a wing-spread of from 29 to 36 inches. This hawk also differs from the sharp-shin in that both the tarsi and the tail are more rounded. The top of the head is also darker than that of the sharp-shin.

These two hawks, the "big blue and little blue darters," are the true "chicken hawks" for they are not only a constant menace to the bird-life and other small creatures of field and forest but are among the greatest enemies of the poultryman, especially where these activities are carried on in the open. Well do I recall the many occasions on the farm when the sudden sallies of this hawk brought a terrified outcry from the hen-yard, and as I rushed out with the shot-gun it was only to catch a fleeting glimpse of a cooper's hawk disappearing through the willows along the creek with a "frying chicken" tightly clutched in its relentless talons. Occasionally, when present on the spot and at the very moment of the sally of one of these hawks into the poultry-yard, I have succeeded by means of a vigorous outcry and clapping of hands in causing the hawk to drop its prey, but this does not often occur. In case a pullet or hen is too large easily to be carried off such fowl is sometimes killed and partially eaten on the spot to which the hawk may return again and again. Even a sizable and formidable rooster has been known to have been killed and left dead on the spot, such is the limitless audacity and killing power of this relentless pirate of the air.

Scientists of the United States Department of Agriculture, some years ago, reported that of 133 stomachs of this hawk which were examined, 34 contained poultry and game birds; 52, other birds; 11, mammals; 1, frog; 3, lizards; 2, insects, and 39 were empty.

This hawk breeds from southern Canada, south to the southern border of the United States. It is resident in Indiana. The nest is well up in trees, of sticks, twigs and bark. The eggs are 3 to 6, pale bluish-white, sometimes faintly spotted with brown.

MARSH HAWK
(Circus hudsonius)

The marsh hawk, as its name indicates, is a bird of the open country. Its favorite haunts are marshes and wet prairies. Here it may be seen hawking about with measured wing-beats and hovering over some especially enticing spot for the small quarry which constitutes the greater portion of its prey. It is sometimes called bog hawk, mouse hawk and frog hawk. The male is from 17½ to 20 inches in length with a wing-spread of from 40 to 45 inches, while the female is from 19 to 24 inches with a wingspread of from 43½ to 54 inches. This hawk easily may be distinguished from other members of the family by the conspicuous white rump which in a good light is clearly visible from a considerable distance as the bird performs its leisurely evolutions over field or marsh.

Although a true hawk, and a skillful hunter of the quarry which constitutes its prey, this hawk seems to be quite lacking in the fierce and aggressive qualities which are characteristic of some other members of the family. For instance an old hen, endowed with something of the fighting spirit in the matter of protecting her family, has been known to drive a marsh hawk away from her brood. Blackbirds and other birds will often chase them and even terns have been known to mob a marsh hawk when in the neighborhood of their breeding grounds. However, the marsh hawk is often quite bold and pugnacious during the breeding season and will attack larger hawks, vultures and even eagles when these birds are in the vicinity of their nest. The courtship of the marsh hawk is of special interest and is carried on in the birds' favorite element, the sky, which during the hours of daylight is its home. During the warm days of spring a pair often may be observed soaring at great height, performing a series of amazing and beautiful evolutions in the air. Tumbling and somersaulting and parachuting and then mounting skyway again, the entire performance arouses the admiration of the most casual observer, for the birds have proved themselves to be perfect masters of the air. After alighting on the meadow or marsh another less spectacular but none the less interesting courtship performance is gone through with.

The nest life of the marsh hawk also has about it much of interest and charm and has been observed and described by various naturalists. The male has proved himself to be a bountiful provider.

In some places and during certain months of the year the marsh hawk has been regarded as much more beneficial than harmful. Of 124 stomachs examined by Dr. A. K. Fisher, 7 contained poultry or game birds; 34, other birds; 57, mice; 22, other mammals; 7, reptiles; 2, frogs; and 8 were empty. This hawk ranges throughout North America. It is resident in northern Indiana. The nest is on the ground in marshes. The eggs are 4 to 6, pale bluish white.

NORTHERN RED-SHOULDERED HAWK
(*Buteo lineatus lineatus*)

Perhaps the average person would think one quite foolish to speak or write with enthusiasm about any member of the hawk family. However this may be, a number of years ago there was organized the "Hawk and Owl Society." The membership was made up of naturalists, both professional and amateur, who were especiallly interested in the so-called birds of prey and who wished to wage both educational and legislative campaigns against their indiscriminate killing.

The red-shouldered hawk is both a conspicuous and interesting member of the family. The male is from 18 to 23 inches in length with a wing-spread of from 33 to 44 inches. The female averages larger than the male. It is always a pleasing sight to see a pair of these great hawks, or the red-tail, their nearest of kin, soaring far overhead in the pale blue sky of April. It serves to add a touch of the wilderness and to remind us that our civilization has not quite yet succeeded in vanquishing the last vestige of our wildlife. The most common note of this hawk is a long drawn out *tee-ur tee-ur*, or *kee-you, kee-you*. It also occasionally utters the note, *cac-cac*, which reminds one somewhat of the bald eagle.

Doubtless most persons when they see a large hawk soaring overhead or perched on the dead limb of a tree, perhaps by the highway, at once pronounce it a "chicken hawk." The red-tail and the red-shoulder are decidedly not "chicken hawks" for this name is appropriately applied only to the sharp-shinned and the cooper's hawks. Very rarely does the red-shouldered raid the poultry yard. Its food consists mostly of small rodents principally mice, also batrachians and snakes. Dr. B. H. Warren, in the "Birds of Pennsylvania", gives the stomach contents of 57 of these hawks as follows: 43 stomachs contained field mice, a few other small mammals and insects; 9, frogs and insects; 2, small birds; 3, remains of small mammals and beetles. Dr. A. K. Fisher author of "Hawks and Owls of the United States" has reported on the stomach contents of 220 of these hawks as follows: Only 3 contained poultry; 12, other birds; 102, mice; 40, other mammals; 20, reptiles; 39, batrachians; 92, insects; 16, spiders; 7, crawfish; 1, earthworms; 2, offal; 3, fish; and 14 were empty.

The results of these studies as here given should serve to convince even the most skeptical as to the economic value of the red-shouldered hawk.

This hawk is found in varying numbers and in its three recognized forms from southern Canada and Prince Edward Island south to the Florida Keys and west to the edge of the Great Plains. It is resident in Indiana although its distribution varies with the season and with different localities. The nest is in a tall tree of sticks and twigs lined with grass. The eggs are 3 to 5 dull white, sprinkled and blotched with chocolate.

18

EASTERN RED-TAILED HAWK
(Buteo borealis borealis)

The red-tailed hawk is the largest of our common hawks, with the exception of the fish hawk or osprey. The male is from 19 inches to 22 inches in length with a wing spread of from 46 to 50 inches while the female measures as much as 25 inches and has a maximum wing spread of 56 inches.

It is quite unfortunate that this fine hawk, like its relative the red-shoulder, should have been called by the wholly inappropriate name of "chicken hawk." However, those who are responsible for our Indiana game laws displayed a fine sense of discrimination and placed both this hawk and the red-shoulder on the list of protected birds. There has long been a deep-rooted and ineradicable prejudice in the minds of the general public, especially farmers, poultrymen and sportsmen, against all of our so called birds of prey. One nationally prominent leader in the field of conservation, and chief of the game commission of his great state, once wrote me that he considered "the only good hawk is a dead hawk," as his reply to my questionnaire which was sent to the game commissions of all the states in the Union.

Many years ago, before we had accumulated the great store of facts we now possess as to the intricate relationships of living things, the State of Pennsylvania, for a period of several years, paid certain bounties to those who killed hawks and owls and delivered the heads as proof of their prowess. After this program had been in effect for a few years the state was visited with an unusual infestation of field mice and other rodent pests. Realizing the mistaken policy the bounty law was repealed. Since that day bounty laws as a means of controlling any kind of predator or pest have been discredited and other more scientific methods have been substituted.

Now as to the food habits of the red tail—Dr. A. K. Fisher, years ago, recorded that of 562 stomach contents examined 54 contained poultry; 51, other birds; 278, mice; 131, other mammals; 47, insects; 37, batrachians or reptilian forms; 8, crawfish; 13, offal and 89 were empty. Other investigators have reported only mice. So it is evident that this hawk shows a wide range in its food habits. Now and then individual hawks may acquire the habit of taking poultry and in such instances effective action should be taken. Forbush has reported that the red-tail is 85 per cent beneficial. Moreover something should be said in its favor on behalf of that increasing number of persons who go afield in quest of nature lore and in the enjoyment of the great out of doors. Our Hoosier scene would be greatly impoverished were this striking and interesting bird to disappear.

The Eastern red-tail is a common resident of Indiana, more abundant in the southern part of the state. The nest is usually high up in trees. The eggs are 2 to 4, white, marked with cinnamon brown.

19

SHARP-SHINNED HAWK
(Accipiter velox velox)

This dashing and intrepid hawk, which is from 10 to 14 inches long, is sometimes called "little blue darter" to distinguish it from its larger relative, the cooper's hawk, which is often called "big blue darter." The name "little blue darter" is very appropriate inasmuch as it serves to call attention to the hunting habits of this little bundle of animated fury. Wholly unlike its larger and more sedate relatives, the Buteos, which are slow and comparatively heavy in their movements, the sharp-shin dashes with great speed from out of some shady nook, or place of concealment, and strikes down and carries off its quarry in the twinkling of an eye. Not only do our domestic fowls know and fear this fierce little predator but the wild birds of field and forest as well. It has been known to overtake and catch a bob-white in full flight which is undeniable proof of its prowess.

In view, therefore, of the destructive character of this hawk and its close relative, the cooper's, the name of "chicken hawk" should have been applied to these rather than to the red-tailed and the red-shouldered. The well known ornithologist Ned Dearborn has recorded this incident which came under his observation. "It's (the sharp-shin's) audacity when hungry is astonishing. I have seen one pounce on a chicken, right in the village, and wait until it had very deliberately fixed its claws in the chicken's back, eyeing at the same time a man, just across the street, with the greatest insolence imaginable. I once saw one of these hawks dash among a flock of goldfinches that were feeding in a weedy run. They took flight precipitately in all directions but he singled out one and gave chase. No matter how that goldfinch turned, the hawk was always headed for his mark, and constantly nearing it. It seemed as if every tack of the little bird was anticipated by its relentless pursuer. I suppose less than a minute after the hawk's appearance he had the goldfinch in his clutches. The final scene was enacted within thirty feet from my face, yet such was the lightning-like quickness of the hawk's grasp, that I could not perceive it."

Edward Howe Forbush relates that one day William Brewster "was standing by the corner of one of his barns, near the opening of the barn cellar, watching a phoebe which had a nest in the cellar and was flycatching from a low limb near by. A sharp-shinned hawk made a sudden rush at the phoebe. The little bird avoided the stroke and then shot down directly at Mr. Brewster, passed under his elbow, around his body and vanished into the cellar, while the hawk, completely baffled, quickly withdrew."

This hawk breeds throughout all temperate and sub-arctic America. The nest is in trees, of sticks lined with bark and leaves. The eggs are 3 to 5, white, greenish or bluish-white spotted with brown.

20

EASTERN SPARROW HAWK
(Falco sparverius sparverius)

This attractive and interesting little falcon is the smallest member of the family to which it belongs, averaging about 10 inches in length. It is not only the smallest member of the family but it generally will be agreed that the male, at least, is the most strikingly adorned of any of the clan. With his chestnut rufous tail, tipped with black and white, and his whitish face marked with two curving black bars he is certainly calculated to attract the attention of all observers. The female is similar in appearance, but much duller. I would be willing to hazard a guess that the sparrow hawk population of Indiana is greater than that of any other raptorial bird.

As one motors along our highways during any summer month these little hawks may be observed perched on the telephone wires or posts from which points of vantage they swoop down in graceful flight to pick up many a beetle or grasshopper on the pavement, or perhaps to pounce upon a field mouse which their keen eyes have spotted in the adjoining field. Unfortunately there long has been a deep rooted prejudice in the popular mind against all the so-called birds of prey and the sparrow hawk has not entirely escaped this widespread antipathy.

Dr. A. K. Fisher, who spent many years making careful studies of the food habits of the hawks and owls of the United States, reported that "of 320 stomachs examined, 1 contained a game bird; 58, other birds; 89, mice; 12, other mammals; 12, reptiles or batrachians; 215, insects; 29, spiders; and 29 were empty." When grasshoppers are abundant these insects seem to constitute the major portion of the sparrow hawk's food. Sometimes during winter when insects are not to be had they feed upon English sparrows and occasionally starlings. Rarely they may visit the poultry yard and carry off young chickens.

Being a true falcon the sparrow hawk is a dauntless little predator and does not hesitate to attack birds fully as large as itself, even a blue jay or a flicker. It is always interesting to watch a sparrow hawk hanging suspended in the air on fluttering wings as its keen eyes gaze downwards to detect some unwary field mouse in its runway in the grass. The fighting spirit and prowess of this little hawk are revealed in rare instances where it has been observed to attack a crow and a red-shouldered hawk when in defense of its young.

Sparrow hawks begin to mate in early April. Often, during courtship, the male rises from the perch and hovering over the female mounts high in the air and after circling about poises on quivering wings before joining her on the perch below.

The nest is in a hole in a tree, often in an abandoned nest of one of the larger woodpeckers. The eggs are white with numerous markings of various shades of brown.

21

KILLDEER
(Oxyechus vociferus vociferus)

The killdeer has the distinction of being the only member of its genus which is found in the United States. Closely allied races are resident in the West Indies and on the coast of Peru. It is doubtful whether any member of the plover family is as well known as this truly handsome and interesting frequenter of stream margins and the shores of lakes and ponds. It is also common in closely grazed upland pastures as well as plowed fields in the early spring where it often runs before the plowman and his team and even today, before the tractor as it turns furrow after furrow. When overtaken or disturbed it takes flight and continues to utter its rather plaintive and far-carrying note of *kill-dee, kill-dee, dee,* or now and then an even more distinct note of *kill-deer.* Often on quiet moonlit nights, when all the countryside is wrapped in slumber, the notes of killdeers may be heard dropping out of the sky. As a farm lad, the killdeer was almost as familiar to me as the robins that came and nested in the trees of our lawn, or the bronzed grackles that nested in the pines. When the work horses were brought in from the fields for the feeding and watering at noon, killdeers were usually present, racing along the margin of the little creek. Then often, as a sturdy plow-boy following the team of horses, I have stopped to watch the strange and frantic efforts of a brooding killdeer as feigning injury, she endeavored to entice me away from the vicinity of the nest which with some difficulty I usually discovered, perhaps, only a few feet distant from the furrow that was being turned. At such times the behavior of the bird was quite ludicrous to behold. She seemed to be saying, "Just look at me and have pity, I can't get away, both my legs and wings are broken." And as you advance she continues to recede with many ridiculous antics and wing-flutterings until she seems satisfied that she has enticed you a sufficient distance from the nest, which as a matter of fact you haven't as yet discovered. Then you begin to make a diligent search and after a while you are rewarded by finding the pear-shaped eggs in a saucer-shaped hollow on the bare ground. During the years of my life on the farm I am sure that no season passed when at least one killdeer's nest was not found at the time of the spring plowing.

The killdeer is about 10½ inches long and is an active and energetic bird and usually in motion unless sitting on the nest. The food of the killdeer consists largely of insects and earthworms. It eats a great variety of such insects as cutworms, wire-worms, beetles and caterpillars. It is usually a winter resident in some numbers in the southern part of the state. It ranges throughout most of North America and parts of South America.

The eggs are usually 4, creamy white and spotted with chocolate brown.

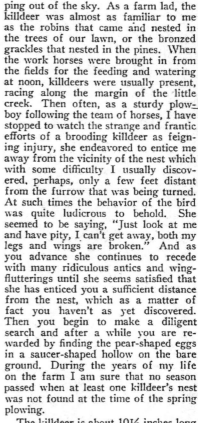

EASTERN BELTED KINGFISHER
(Megaceryle alcyon alcyon)

It is to be doubted whether anyone of our native birds is more eccentric both in appearance and habits than the kingfisher. He is not only "tousled headed" but he has the strange habit of plunging into a stream or lake or pond for the greater portion of his food, and then last of all, when he decides to go to housekeeping he excavates a tunnel in the bank of a stream, lake or pond. Sometimes this excavation, which is about 4 inches in diameter, runs back ten or more feet although it usually is from four to five feet in length. He is above all the one bird who is the patroller of our streams and few creeks, rivers and even smaller water courses within the state, are without kingfishers. He may be regarded as a believer in what modern ornithologists call territorialism for he seems in most instances to have worked out a mutual understanding with the various members of his clan under which both he and his mate carry on their piscatorial calling within a restricted area. This patroller of streams loves to perch on some limb over the water and, although apparently in deep meditation, his keen and watchful eye suddenly descries some finny prey when down he goes like a flash to emerge a few seconds later with the minnow in his stout beak. This he proceeds to swallow and during the process he constantly raises and lowers his crest.

The kingfisher when launching from his perch for a flight down the stream usually utters a series of long harsh rattling notes, quite difficult to describe, although some one has said they resemble "the sound of a watchman's rattle". Maurice Thompson in his poem, "The Kingfisher", has given us a good description of the bird, the first stanza of which is as follows—

"He laughs by the summer stream
Where the lilies nod and dream,
  As through the sheen of water cool and
    clear
  He sees the chub and sunfish cutting
    sheer."

I like our poet's use of the word "laughs" for above all the notes of the kingfisher impress me as being a sort of weird ironic laughter. William L. Finley has described this bird in part as follows—

"A young kingfisher seems to grow like a potato in a cellar, all the growth going to the end nearer the light. He sits looking out toward the door and of course his face naturally goes to nose. Everything is forfeited to furnish him with a big head, a spear-pointed bill, and a pair of strong wings to give this arrow-shaped bird a good start when he dives for fish."

The eastern belted kingfisher, which is about 13 inches in length, breeds from southern Labrador and Manitoba south to Florida. It winters in Virginia, southern Indiana, southern Illinois southward. The nest is usually at the end of a tunnel in an eastern bank. From 5 to 8 glossy white eggs are laid.

PRAIRIE HORNED LARK
(Octocoris alpestris praticola)

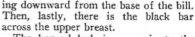

Although a fairly common resident in the northern portion of the state, the prairie horned lark is much more in evidence during the winter months. At this time it has a flocking habit and often appears about the feed-lots of farms where it picks up a portion of the grain left from the feeding of live stock. Often I have observed them on bitterly cold winter days as they have come wheeling across the sky in small flocks to light on some wind-swept hill of a pasture, or perhaps suddenly to swerve in their erratic flight to land in some more enticing spot.

The horned lark is a bird of the open country and it is doubtful if one ever lights in a tree. One well-known ornithologist has stated that he has never seen it perch even on a rail or fence post. It is a bird of rather striking appearance so far as the facial and head markings are concerned and should always be fairly easy to identify even by the amateur bird student. First there are the two "horns" on the sides of the crown, then the black "mustache" curv-

ing downward from the base of the bill. Then, lastly, there is the black bar across the upper breast.

The horned lark is a cousin to the famous skylark of Europe and even though his musical performance is perhaps not worthy to be compared to that renowned bird, a number of ornithologists have waxed eloquent in praise of its song. Among these is Bradford Torrey who writes in part as follows, "His method at such times was a surprise to me. He starts from the ground silently, with no appearance of lyrical excitement, and his flight at first is low, precisely as if he were going to the next field. Soon, however, he begins to mount, beating the air with quick strokes and then shutting his wings against his sides and forcing himself upward, 'diving upward' were the words I found myself using. Up he goes, up, up, up, higher and higher—'till after awhile he breaks into voice. While singing he holds his wings motionless, stiffly outstretched, and his tail widely spread, as if he were doing his utmost to transform himself into a parachute—as no doubt he is. Then, the brief hurried strain delivered, he beats the air again and makes another shoot heavenward. . . . At last he came down; and this, my friend and I agreed, was the most exciting moment of all. He closed his wings and literally shot to the ground like an arrow. 'Wonderful', said I, 'wonderful'."

Horned larks appear to be one of nature's great successes for they are extremely hardy and adaptable and in varying forms are distributed widely over the continent. The American Ornithologists' Union has recognized sixteen members of the genus. The prairie horned lark, which is from 6¾ to 7½ inches in length, is distributed throughout eastern North America, from Maine and Ontario to Kansas, Missouri and Indiana. The nest is on the ground of grass, lined with feathers and other soft materials. The eggs are 3 to 5 grayish-white blotched and speckled with brown.

EASTERN MEADOWLARK
(*Sturnella magna magna*)

The meadowlark is a common summer resident throughout Indiana and usually winters in considerable numbers in the southern portion. In addition to our eastern meadowlark which is from nine and one-half to eleven inches in length, two slightly varying forms are recognized by the American Ornithologists' Union, southern meadowlark and Rio Grande meadowlark.

With most people in the northern states the meadowlark is a highly esteemed bird both for practical and sentimental reasons. The greater portion of its food during the summer months consists of insects most of which are destructive to farm, garden and fruit crops. A well known authority in the field of economic ornithology has reported that ninety per cent of the meadowlark's food, at this season, consists of insects nearly all of which are a menace to agriculture and horticulture.

The writer has in his possession a photograph taken by Dr. Arthur A. Allen of Cornell University. The camera caught the bird with its beak full to overflowing with grub worms which it had taken from a badly infested lawn where the grass was being destroyed. These were being taken to the young in a neighboring meadow where the nest was nicely concealed. When one considers that this particular trip was only one of many which took place during the day, and for many days as well during which the young remained in the nest, some idea can be formed as to the usefulness of the meadowlark. It eats all the principal insect pests of field and garden. It is especially fond of cutworms, caterpillars and grasshoppers. Even during winter months, one authority on the food habits of birds, reported that about one-fourth of the meadowlark's food consists of insects. These studies, therefore, show beyond any doubt that this bird is one of the farmer's best friends. However, we would be quite unappreciative if we were not alive to its esthetic and sentimental value.

The song is a series of clear flute-like notes which when heard floating over the greening fields of spring brings to all who understand a message of optimism and good cheer. To me it has always seemed to say, "*spring o' year,*" or "*spring is here.*" Some one has interpreted its song as, "*Laziness will kill you.*" To me the song, although flute-like, has a penetrating, far-carrying quality about it and often as I motor along over Indiana highways on a sunny, chill March day with the car windows closed, the silvery notes reach me from many a greening pasture or field of young wheat.

The nest of the meadowlark is on the ground usually in a pasture or meadow, and generally quite well concealed. It is constructed almost entirely of dead grasses and sometimes lined with horse hair. From 3 to 5 eggs are laid. These are white, speckled and blotched with brown and lavender.

25

EASTERN-MOCKING BIRD
(*Mimus polyglottos polyglottos*)

One fine, warm April day some two years ago the writer, as guest of the superintendent of schools of Orange county, had just visited a small school in a rather remote part of that charming and picturesque hill country.

A talk had been given on the general subject of the great need for the conservation of our natural resources, especially stressing the value of birds to agriculture and horticulture. As we were leaving the school we were greeted by a medley of bird notes. However, easily outranking them all in richness and continuity of melody was the song of a mockingbird. He was perched on the topmost dead limb of a giant oak which stood on a corner of the school grounds. Now, of course, the bird we saw and heard was the *true* mockingbird, which is generally associated in our minds with the south, with orange groves and magnolias and live oaks. Our more familiar brown thrasher is sometimes called the "northern mockingbird" but it should never be confused with its rival.

Most Hoosier bird students are familiar with the fact that the mockingbird, during recent years, has gradually been extending its northward range not only within our state but also in other parts of the country.

Edward Howe Forbush, long state ornithologist of Massachusetts, has recorded on a map the rather wide winter distribution of the bird in that New England state. One zero morning two years ago, when much snow and ice covered both ground and trees the birds naturally were in a sorry plight. We repeatedly watched a mockingbird at our home in Mooresville as it bravely tried to withstand and overcome this trying situation. Two or three mornings in succession it came to our front porch and ate the seeds of both dogwood and sumac from a "winter bouquet" which had been put in a stone jar in the fall.

Mention was made in the beginning of the song of the mockingbird. It is a famous and versatile vocalist and occupies much the same position in our country that the nightingale does in Europe. I think it will be agreed that it takes first place among all our songsters, although there may be those who would regard the brown thrasher as a close competitor. Whether its song be heard on a still, moonlight night amidst the orange groves of Florida or on a bright May morning in the Hoosier state, the melody carries with it an indescribable charm. Many nature writers and poets, including our own Maurice Thompson, have vied with one another in praising and attempting to describe the vocal performance of this bird.

It may seem a bit strange that a bird with such a refined and classic song should be a bit pugnacious. He is the boss about a feeding station and will even attack a cat. He is about 10½ inches long.

The nest is placed in a variety of situations. Four to 6 pale greenish-blue eggs, spotted heavily with brown, are laid.

WHITE-BREASTED NUTHATCH
(Sitta carolinensis)

To many bird students the white-breasted nuthatch is a bird of autumn days and bare and leafless winter woodlands, and doubtless it is more in evidence during these seasons. This little patroller of tree trunks, which is from 5¼ to 6¼ inches in length, is one of nature's feathered acrobats, for it is equally at home as it climbs head downwards on a tree trunk or as it proceeds upward round and round the bole. As it progresses it constantly utters a muffled nasal *hank, hank,* or *yank, yank,* which is audible for only a short distance. Bird notes and songs are always difficult of being adequately rendered in words or syllables as they mean different things to different individuals. Dr. Frank M. Chapman has stated that in the spring, during the mating season, the monotonous call note is varied by a "tenor *hah, hah, hah*—sounding strangely like mirthless laughter."

It was stated above that the nuthatch is one of nature's feathered acrobats. Edward Howe Forbush has written in his "Birds of Massachusetts":

"No other bird can compete with the nuthatches in running up and down a tree trunk. They are so often seen creeping head downwards that some country people call them "Devil-down-heads", or "Upside-down-birds". They seem to have taken lessons from the squirrel which runs down the tree head first, stretching out his hind feet backward and so clinging to the bark with his claws as he goes down; but the nuthatch, having only two feet, has to reach forward under its breast with one and back beside its tail with the other, and thus, standing on a wide base and holding safely to the bark with the three fore claws of the upper foot turned backward it hitches nimbly down the tree head first—and it runs around the trunk in the same way with feet wide apart."

The mating season and nest-building period, as in the case of many birds, is attended with interesting activities. The male is a very gallant suitor and in ad-dition to ruffing up his feathers and partly spreading his wings in an attempt to impress his mate with his many and outstanding attractions, he often manifests his ardor in a more practical way. For instance he will even shell seeds for her and then pass her the hulled kernels.

The nuthatch is of much value as a destroyer of insect pests. It is recorded that the stomach of one bird contained 1629 eggs of the fall canker-worm moth. The larvae of the gypsy moth and tent caterpillar are also eaten.

In addition to the type species of the white-breasted nuthatch, six sub-species are recognized by the American Ornithologist's Union. In Eastern U. S. we have the white-breasted nuthatch and the Florida nuthatch. The nest is always in a hole or cavity and may be from two to sixty feet up. It is lined with fine grasses, bits of fur, shreds of bark, etc. From 5 to 8 eggs, creamy white and spotted with reddish brown, are laid.

27

BARN OWL
(*Tyto alba pratincola*)

This owl, by reason of its truly remarkable and peculiar facial disc, has come to be widely known as "monkey faced owl". Although not evenly distributed throughout its range, and usually not common anywhere, it always makes good copy for the local newspaper when one of the birds has been reported, or perhaps wantonly shot through ignorance of its great economic value, or of the state conservation law under which it is protected. The barn owl is from 15 to 21 inches in length with a wing-spread of from 43 to 47 inches. The female usually averages somewhat larger than the male, but sometimes is smaller. I think most persons who are acquainted with this owl will agree that the color pattern, even without the striking facial expression, gives a touch of distinction to this elusive and unusually interesting member of the family. The color above is a combination of the most delicate shades of

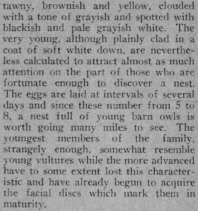

tawny, brownish and yellow, clouded with a tone of grayish and spotted with blackish and pale grayish white. The very young, although plainly clad in a coat of soft white down, are nevertheless calculated to attract almost as much attention on the part of those who are fortunate enough to discover a nest. The eggs are laid at intervals of several days and since these number from 5 to 8, a nest full of young barn owls is worth going many miles to see. The youngest members of the family, strangely enough, somewhat resemble young vultures while the more advanced have to some extent lost this characteristic and have already begun to acquire the facial discs which mark them in maturity.

The barn owl, perhaps to a greater extent than some other members of the family, scorns the light of day and may be found in the half gloom of a barn loft or some abandoned house, or perhaps in the hollow of a tree or in a cave.

One night in early December as I was motoring to Clifty Falls State Park, I observed one of these owls in the lane of light just ahead of my car as it flew noiselessly across the highway. The bird was evidently out for a night's hunting, in neighboring fields.

Two Boy Scouts in Lawrenceburg informed me, recently, that they knew an abandoned house which was occupied by 6 or 8 of these owls. I have met a number of farm-lads throughout the state who have told me that a pair of barn owls were welcome visitors in their barns for these lads were acquainted with the beneficial habits of this owl. Of all the members of the family the barn owl is recognized as the most useful for a large percentage of its food consists of small rodents. It ranges from southern Canada over most of the United States and south to Mexico and Central America.

The nest is almost anywhere, deserted buildings, towers, and holes in trees or banks. The eggs are 5 to 8 and chalky white.

NORTHERN BARRED OWL
(Strix varia varia)

One fine day in October some years ago my itinerary took me to a small elementary school in the heavily forested section of a southern Indiana county. The purpose of my visit was to talk to the pupils on the general theme of the great need for the conservation of our natural resources, especially of our forests, our fish and game and bird-life. In the course of my talk I asked a number of questions in order to learn something about the extent of the nature lore, which might be stored in the minds of this fine group of youngsters. A good part of our discussion was about birds and their value and the laws protecting them. We came to the subject of owls and one particularly bright sixth grade boy informed us that two kinds of "hoot owls" were quite common in that section. However, he said that these owls had two quite different "hoots" and then upon being asked to imitate their "hootings" he instantly consented by giving us an amazing exhibition as an imitator of bird notes.

As is well-known to all bird-watchers in our state two kinds of "hoot owls" are common residents in the more heavily forested sections, especially in our state parks such as Brown County, McCormicks Creek and Spring Mill. They are the great horned and the barred owl. The former "hoots" in a monotone uttering a series of monotonous *Whoo-hoo-whoo-hoos* which may be heard for a mile or more on still nights. The notes of the barred owl, however, are full of amazing modulations reminding one of a sort of weird demoniacal laughter. Often when camping alone in some remote cypress swamp in the wilds of Florida, I have been serenaded night after night by the eerie music of numbers of these somber visaged birds. They seemed to be endlessly interrogating me — *"Whoo-whoo-whoo-whoo—who* are YOU all?"*, the emphasis being placed on the final *you*. I have always experienced a certain thrill and pleasure from listening to the strange, weird music of the barred owl

and I would journey miles just to enjoy a midnight serenade. Not only may they be heard of nights but often of dull cloudy days their hootings resound throughout the forest aisles.

The barred owl is regarded as on the whole a beneficial bird and as such it has legal protection under state law. Although a powerful bird and capable of causing damage to poultry and small game, it has been found to feed upon mice and other small mammals, also frogs, lizards and crawfish as well as large insects. The northern barred owl which is from 19 to 24 inches in length breeds from northern Ontario, Manitoba, southern Quebec and Newfoundland south to Tennesssee and northern Georgia and west to Montana and Colorado.

The nest is usually in a hollow tree or sometimes in the deserted nest of a hawk or crow. Two to 4 pure white eggs are laid.

GREAT HORNED OWL
(Bubo virginianus virginianus)

The great horned owl, which is nearly 2 feet in length, by reason of its size and prowess as a skillful hunter of many kinds of prey, is a conspicuous member of the large family to which it belongs. In its eight recognized forms it ranges widely over the continent from the Canadian Provinces and Alaska south into Mexico. However little or however much it may vary in size or appearance it ever remains the same fierce and formidable member of the owl clan. Just now as I pen these lines there comes back the vivid and unforgettable memory of an experience I had with one of these birds.

A neighbor had shot one which had been catching his chickens. It was uninjured save for a broken wing. While in the process of putting the beak into a cyanide of potassium bottle as a humane method of killing the bird, one foot armed with four cruel claws suddenly gripped me by the wrist. It was only with help and the use of two pairs of wire pliers that I was able to free myself from the talons of this powerful bird.

On account of the fierce and morose character of this owl it has few if any friends and it is not protected under the laws of any state. If, as was more frequent in former years, poultry is permitted to roost in the trees about the premises this owl often makes heavy inroads on a flock. A school teacher in a rural school in Orange County informed me, not long ago, that these owls had taken between 65 and 70 of their chickens which were roosting in some apple trees. Amos Butler in his "Birds of Indiana" reports E. J. Chansler as having informed him that one fall he lost 59 young guineas by the depredations of these owls. Edward Howe Forbush has stated in his book "Birds of Massachusetts" that, "Every living thing above ground in the woods on winter nights pays tribute to the great horned owl except the larger mammals and man." He then relates the authentic incident of a large tom cat having been caught in the clutches of one of these owls, on a moonlit night, and having "its vitals torn out" before the owner could get his gun.

Enough has been said to portray the fierce and predatory nature of this skillful night hunter. Herber L. Stoddard, the well-known ornithologist and wildlife technician who has spent many years on the extensive quail preserves of southern Georgia and northern Florida, has found that the great horned owl is the best friend the quail has under the particular conditions existing in that region. The large population of cotten rats is the single greatest menace to the quail, since a large percentage of the eggs are destroyed by the rats. Strange to relate the greatest enemy of, and check on the rats, is the great horned owl which feeds extensively on these rodents.

The nest of this owl is usually well up in a tree, generally in the old nest of a crow, squirrel or hawk. Two or 3 white eggs are laid.

LONG-EARED OWL
(Asio wilsonianus)

I shall always remember the thrill I experienced when many years ago I made my first acquaintance with the long-eared owl. Its really striking appearance, due chiefly to the unusually long ear-like tufts margined with white and buff, and then the handsome facial disc of tawny and black and white, made an impression which has lasted through the years. This owl measures from 13 to 16 inches in length, with a wing-spread of from 36 to 42 inches. The sexes look alike, although the female averages a little larger than the male. The downy young are white and in the first winter plumage closely resemble the adults but are somewhat more rufous.

The long-eared owl is more nocturnal in its habits than some other members of the family. Its favorite haunts during the day are the gloomy depths of coniferous forests, or where these do not occur it often finds a safe and congenial retreat in a clump of evergreens in your door-yard, or even in the barn-loft or in the half gloom of some other farm building. On account of its almost strictly nocturnal habits this species is probably less known than the other owls which are resident within the state, and it is doubtless more common than is generally believed, even by competent bird students. Although a resident species the long-eared owl population, during the late fall and winter months, is greatly increased by migrants from the north.

Forbush in his "Birds of Massachusetts", writing of the long-eared owl says, "The wings and tail of the long-eared owl are so long that in flight the bird looks larger than it actually is. Its body is slight and slim and its feathered feet are small. In winter numbers of these birds are accustomed to hunt for field mice along river bottoms in open country where there are few coniferous trees and to assemble to roost at night in some clump of ornamental evergreens planted to decorate the grounds of some residence. When in danger of being observed in such a retreat they draw the feathers close to the body and stand erect so as to resemble the stub of a broken branch."

Ornithologists have given various accounts of the voice of this owl. O. W. Knight says it is "a peculiar whining cry much like a very young puppy," while G. Clyde Fisher describes it as "a softly whistled *whee-you*, the two syllables slurred together."

It is unfortunate that there should exist a general prejudice in the minds of the public against all hawks and owls, since this, and most other owls are beneficial.

The food of this owl consists largely of mice and other small mammals, also of various beetles and other insects. Its range is throughout most of North America. The nest is usually in an old crow's, hawk's, or squirrel's nest. From 3 to 6 white eggs are laid.

31

EASTERN SCREECH OWL
(Otus asio naevius)

Among the owls of North America the little screech owl, in its varying forms, is doubtless best known to the general public. Perhaps it has learned to appreciate the advantages of human neighbors, for it comes confidingly and makes its home in our barns and other outbuildings until the nesting time approaches when it may seek some natural cavity in an old apple tree in the nearby orchard, or perhaps in the decayed crown of an ancient beech in the lawn by the farm house. Like a number of other birds which have been included in this series, the screech owl is one of nature's great successes for it seems able to adapt its mode of life to all sorts of conditions. It occurs in 15 varying forms throughout the continent, from Alaska to Mexico and from Maine to Florida.

Our eastern screech owl, the type species, varies in length from 6½ to 10 inches, with a wing-spread of from 18 to 24 inches. This little owl has the distinction of possessing two entirely different color phases which have no relation to age, sex or season. In one phase the general color tone is bright rufous and in the other, grayish. In his "Birds of Indiana" Amos Butler quotes Robert Ridgway as reporting that 95 per cent of all the screech owls in the Wabash Valley were found to be red, whereas Butler reports that in the White Water Valley only 60 per cent were red. Other records are available which show a wide variation in the percentages of the two color phases.

In spite of its comparatively small size and innocent appearance this little owl, at times, is a veritable whirlwind of spiteful audacity when its young have just left the nest and are perched in the shade trees of your lawn. Well do I recall the many occasions, when the parent birds apparently resenting our presence, as in the gathering dusk we sat near the verandah, made repeated and spirited attacks striking us with their sharp claws on our bare heads, first one and then another member of the family being the victim. At first one was inclined to laugh at these twilight sallies of this little feathered gladiator, but as a matter of fact if repeated several times we usually ended by seeking the shelter of the living room.

Others have reported similar experiences in being attacked by this little owl. These attacks are by no means laughing matters for the sharp claws often penetrate deeply into the scalp, and are really quite painful. These spirited advances give evidence of the prowess of this little predator.

The screech owl is regarded as more beneficial than harmful. Of 255 stomachs examined only one contained poultry. It eats mice and other small rodents as well as large numbers of insects. The nest is in a hole in a dead tree or stub, often in a nesting box. From 4 to 6 white eggs are laid.

SHORT-EARED OWL
(Asio flammeus flammeus)

The short-eared owl, unlike its nearest of kin the long-eared which is a forest inhabiting species, is a bird of marsh lands and grassy prairies. Like the burrowing owl it has departed from the usual habits of the clan and has taken to the wide open spaces. The length is from 12½ to 17 inches, with a wing-spread of from 38 to 44 inches. The female is usually somewhat larger than the male which she closely resembles. The ear-tufts of this owl are small and very short. The conspicuous whitish markings on the face add a touch of distinction and portray its close kinship with the long-eared owl. My first acquaintance with this owl was made years ago on the Indiana farm when as a fall and winter visitant they came from their summer haunts far to the northward. Then some years later I met them on the wet prairies and savannas of the south. It is sometimes called marsh or bog owl on account of such regions being favorite hunting grounds.

The short-eared owl is to a great extent a hunter by day and is wide-ranging in its favorite haunts, and both during the breeding season and during migration is often found in salt marshes along our sea coasts as well as along the grass-covered sand dunes and on off shore islands. In such situations it not only feeds upon rodents but also upon a great variety of the larger insects which it often catches as it flies either over marsh or over the water. Its flight somewhat resembles that of the marsh-hawk for it is characterized by an easy grace and seems to be effortless. Forbush who had an intimate acquaintance with this owl wrote as follows: "When it sees a favorable opportunity, it may hover for a moment, or may drop directly on its prey and remain there to devour it; or if flying low it may snatch the unlucky victim from the ground, and pass on without even checking its speed, so swift and skillful is the stroke. If 3 or 4 are cruising about together and one stops to kill and eat a mouse, the others are likely to alight also, either on the ground or on nearby bushes as if curious to see what their neighbor is doing."

The mating call of this owl has been described as a series of notes like *toot, toot, toot,* repeated as many as 15 to 20 times.

This owl is of great value when there occur periods of infestations of field mice. At such times they congregate in unusual numbers feeding upon these rodent pests. There are few records of its attacking poultry. It occasionally eats a few young rabbits.

Often of still nights I have been charmed as I have listened to the really mellow musical notes of this owl as they have come floating across some lonely marsh.

The food consists mostly of mice and shrews and various insects. It ranges widely over the continent and is resident in northern Indiana. The nest is on the ground, of sticks and grasses. From 3 to 5 white eggs are laid.

HUNGARIAN PARTRIDGE
(Perdix perdix perdix)

This interesting and popular member of the quail family, which is also called European partridge, Bohemian partridge, English partridge and gray partridge, was first introduced into the United States in 1900 when a number were released in the Willamette Valley of Oregon. In 1904 another importation was liberated in South Carolina and another in British Columbia. During succeeding years many of the birds were imported and released in various states and also in Western Canada. This game bird was first introduced into Indiana in 1907 and by 1910 a total of 12,000 birds had been imported and released in every county of the state. No important releases have been made since these first importations.

Wm. B. Barnes, of the Pittman-Robertson Wildlife Research Project, reports as follows concerning the results of these releases: "From this original release this native of Europe has been able to establish its range in a restricted section of the state, thus indicating it is limited by farming practices and soil

types to certain conditions necessary for its existence. The bird appears to hold its own in the clay soils. It does not appear to favor the Kankakee sands, or prairie soils of the northwestern and western part of the state. The bird has also vanished from most parts of southern Indiana where it was originally released in equal numbers." Edward Howe Forbush, State Ornithologist of Massachusetts, reported in 1927 that this "game bird in New England has not proved a great success." Many nests and eggs each year were destroyed by the mowing of the meadows and grass fields. Its failure to seek the cover of woodlots and thickets no doubt is a limiting factor in the survival ability of this game bird.

Like our bob-white the Hungarian partridge roosts on the ground with all the members of a covey forming a circle with tails together and heads out. The male is a very solicitous parent and takes his turn at brooding the young, and both parents attempt to allure any enemy away by their cries and other behavior. The plumage of this game bird, even to a greater extent than that of the bob-white, renders it next to invisable when, in case of alarm or danger, it crouches close to the ground.

Forbush writes of this bird: "The parents and the young keep together during the autumn and winter, and so long as they can find food they remain in the same locality, moving only in case of necessity. In spring their food consists largely of insects and succulent green vegetation. Many insects including ants, plant lice, and beetles are taken, ants being a favorite food."

This partridge is about half way between the size of the ruffed grouse and the bob-white and is from 12 to 14 inches in length with a wing-spread of from 18 to 22 inches.

The nest is a hollow on the ground, in fields, meadows, or waste lands, and lined with grasses and dead leaves. The eggs are 9 to 20, ovate in shape and olive in color.

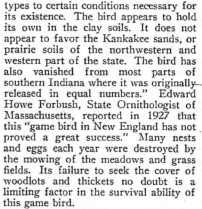

RING-NECKED PHEASANT
(Phasianus colchicus torquatus)

The native home of this attractive and exotic game bird is Asia, that vast region which is so rich in various species of the pheasant family. This group of birds has been given notable treatment by Dr. William Beebe in his monumental work, "Pheasants, Their Lives and Homes."

The introduction of pheasants into Europe goes back to a very remote date and it is thought that the so-called English pheasant became acclimated in Britain in 1299, during the reign of Edward I, the ring-neck having been introduced somewhat later. The first successful introduction of the ring-neck pheasant into the United States was made by Judge O. W. Denny, the Consul General at Shanghai, China, in 1881, when he sent 28 of the birds from China to Oregon. These birds are said to have been a nearly perfect strain, and they soon became well acclimated and increased in numbers.

The ring-neck was first introduced into Indiana in 1899 by the Commissioner of Fisheries and Game, Z. T. Sweeney. He imported 250 of the birds which were released in small flocks in various parts of the state. Much optimism was shown concerning the project. The birds did show some increase but it soon became evident that this was not sufficient to justify an open season.

During the years that have elapsed since this first introduction the ring-neck has undergone varying fortunes. There have been many ups and downs and the results, perhaps, have justified neither a feeling of optimism nor undue pessimism. However, it is significant that, during recent years, when our ardent Hoosier sportsmen wish pheasant shooting "unlimited" they hie away to the Dakotas where for some reason, perhaps not wholly understood in the ecology of things, this popular game bird finds a happy home.

Those especially interested in the history of the ring-neck in Indiana may obtain the complete picture by examining the records in the files of the Indiana Department of Conservation. Although conflicting opinions may be held concerning the causes of the failure of the ring-neck to become established within the state in numbers sufficient to assure an annual take during an open season, it generally will be agreed that artificial propagation is not the answer to the problem. It would seem that the provision and maintenance of a congenial habitat would be far more important. In 1947 the Conservation Department, using its discretionary power, deemed it advisable to close the season on pheasants.

The male ring-neck is strikingly adorned with its gorgeous facial and neck markings. The white ring around his neck is a conspicuous field mark.

The male ring-neck averages 35 inches in length while the female is about 20 inches. The weight of the male is from 2½ to 4½ pounds. The nest is on the ground. The eggs are 6 to 12, pale olive-brown to pale bluish-green.

637894

GREATER PRAIRIE CHICKEN
(Tympanuchus cupido americanus)

One chill March day in 1930 it was my privilege to witness an event I shall long remember. As guest of the State Ornithologist of Massachusetts, Dr. John B. May, I stood in a field on Martha's Vineyard Island and saw "the last heath hen," the sole survivor of its race, come to its ancestral "booming ground" and there go through its amazing nuptial performance. All efforts on the part of various interested persons and agencies to preserve the last remnants of this once common bird were doomed to failure. The heath hen, as a matter of fact, was just a slightly varying eastern form of our prairie chicken, or pinnated grouse. Now some of us are wondering whether the fate that overtook the heath hen is, in the not very distant future, in store for the prairie chicken, not only within the state but throughout the entire range where its numbers have been rapidly declining, due to the constantly increasing encroachments of civilization. We are indebted to Wm. B. Barnes of the Pittman-Robertson Research Project for the most recent report on the present status of this fine game bird within the state. "The decrease of this species has been surprisingly rapid during the past thirty years. The report of George W. Miles, Indiana Commissioner of Fisheries and Game, in 1912 showed that 28 counties reported having these birds and that there were at least 100,000 in the state at that time." Again to quote Mr. Barnes' report; "In 1941 there were at least 1,000 prairie chickens in the state. Each year since that time a decrease has occurred with a resultant estimated population of 470 birds in the spring of 1945. Nine counties are represented by known booming grounds and four others still contain a few birds . . . the last remaining habitat exists in Newton county. Eleven of the 27 known booming grounds are located in one township. It is here that we are making an effort to preserve a remnant of the native prairie chicken population." So much for the present status of the prairie chicken in Indiana.

The steadily declining fortunes of this once abundant and popular game bird is mute evidence of what has been happening and is no doubt yet to happen, to a few other of our native wild birds and mammals. Some one has said that civilized man is the most destructive creature that has ever occupied the earth. So, however much we may regret these facts, they must be faced with resignation.

This famous member of the grouse family is from 18 to 19 inches in length. The mating performance has been described at much length by various observers, the so-called booming sound being produced as the air is drawn into the sacs of loose, naked skin on the sides of the neck. The original home of the prairie chicken was the prairies of the Mississippi Valley, south to Louisiana, and north to South Dakota, west to Kansas and east to northern Indiana.

The nest is on the ground. The eggs are 11 to 14, pale cream, olive-buff or light brown.

EASTERN BOBWHITE
(Colinus virginianus virginianus)

One fine day in May a few years ago, while motoring over the paved highway through an unusually attractive farming district, there suddenly appeared by the roadside a pair of bobwhites with a sizeable family of newly hatched young. The parent birds were in the act of leading their little family across the highway into a field on the opposite side. I came upon them while in this act and in order to avoid running over them I slammed on the brakes stopping the car within a few feet of the startled birds. The parent birds in a few seconds had crossed the pavement apparently awaiting developments, as the young birds instantly "froze", squatting low and remaining immobile. It was a well-known and common enough bit of acting on the part of this bobwhite family. However, the fine day and the particular setting made an indelible picture which will never fade from my memory. I cautiously got out of the car in order to secure my camera which was in the trunk in the rear, when to my keen disappointment, just as I was advancing to make an exposure, the young birds arose and went skittering across the highway to join their distressed parents who were awaiting them in the grass by the roadside. Many bulletins and books have been written about the bobwhite quail and he needs no introduction to the general public, at least to those of us who are not confirmed city-dwellers. Who is there among us who does not find special enjoyment in hearing his cheery call come floating across the greening fields of spring or again over the frosty ones of autumn? However, our well-loved bird since the settlement of the country, has been the victim of varying fortunes. The two principal limiting factors in his survival ability have been over-shooting and the occasional severe winters. In order to increase the bobwhite population, many game commissions have resorted to both artificial propagation and to restocking by importations, often of birds from Mexico. The value of these methods in bringing about an increase in the quail population has been questioned by many, some of whom maintain that the most important factors are the improvement of habitat, the control of predators and the wise fixing of the open season and bag-limits.

Aside from the value of the bobwhite as a game-bird and the sentimental and esthetic value which we attach to it just as a part of our great out-of-doors, it should not be forgotten that few birds are of greater value to agriculture and to horticulture. Among insects eaten are potato beetles, grasshoppers, locusts, chinch bugs, May beetles, cucumber beetles and many others. It also eats many kinds of weed seeds.

Our Eastern bobwhite, which is from 9½ to 10½ inches long, ranges from Southern Ontario, southern Maine, south to the Gulf Coast and to eastern Colorado. The nest is on the ground, of grasses. From 10 to 25 white eggs are laid.

EASTERN ROBIN
(*Turdus migratorius migratorius*)

Of all the wild birds of our continent the robin is no doubt one of nature's greatest successes. Hardy and resourceful it adapts its way of life to all sorts of conditions and in its varying forms ranges widely over the United States and Canada.

Professional ornithologists, as well as amateur bird students, have made detailed and intimate studies of its life history and behavior.

In most sections of the North it is highly regarded both for its song and for its confiding disposition which leads it to come and make its home in our dooryards. In parts of the South, however, especially in Florida, where it winters in countless numbers, it is in great disrepute on account of its habit of feasting upon the strawberries which ripen in midwinter, during the period of its sojourn. Inasmuch as the growers would suffer great financial losses, even to the point of being ruined, the Federal Government has granted permits to shoot robins. On one occasion I saw in the heart of the strawberry section, an elderly woman with a long bamboo pole with which she was vigorously and frantically beating the corrugated iron roof of a low shack in order to frighten away the great numbers of robins which were flocking to the field to feast upon this valuable crop which was no doubt the principal, if not the only source of income, of the woman and her family.

Many years ago before the enactment of state and Federal laws to protect our wild bird life it was a common practice in many parts of the south to kill robins for food. In an attempt to combat this killing the National Audubon Society launched an educational program in the schools of Virginia. This was the beginning of the Junior Audubon Club movement which has grown by leaps and bounds until more than 7,000,000 members have been enrolled. Happily as the years have gone by public sentiment has been vastly changed and it is only rarely that instances are reported of illegal killing of robins.

It must not be thought, in view of the occasional and regional damage the robin does by eating cultivated fruit crops, that it does not have any value as a destroyer of insect pests. It eats large numbers of grasshoppers, locusts, crickets, wire worms, tussock moths, leaf beetles, cut-worms, army-worms, coddling moths and caterpillars as well as many other harmful insects.

The robin's song is a loud and musical carol which might be rendered *cheer up, cheer up, cheer up*.

Rather than migrate southward robins often brave the cold and snow and ice of winter and there are doubtless few winters during which it may not be observed in the southern part of the state. Our eastern robin, which is from 9 to 10 inches in length, ranges from Alaska, Manitoba and Quebec south to New England and the Ohio Valley. The nest is usually in a tree, of twigs, grass, strings, fibres, and lined with mud. The eggs are 4 to 6 and bluish green.

ENGLISH SPARROW
(Passer domesticus domesticus)

It may be something of a surprise to the average bird-lover to be informed that the ubiquitous and self sufficient little English sparrow is not a sparrow at all, for the American Ornithologists' Union has placed him in the family of weaver finches, the Ploceidae. At any rate whatever may be the particular niche he occupies in the bird world this aggressive little alien, which will average about 6¼ inches long, is here to stay. Like Booker T. Washington's negro he came by "special invitation", and finding a congenial environment everywhere, he has made himself at home throughout the length and breadth of the land.

In the year 1850 eight pairs were introduced at Brooklyn, New York by Hon. Nicholas Pike and other directors of the Brooklyn Museum. Strange to relate the birds did not thrive and two years later many more were imported and confined. Later these were released and their progeny spread over the country. There were other importations during the next decade or so but doubtless the importation of 1852 was sufficient to account for the comparatively rapid spread throughout the country. By the year 1875 they are said to have crossed the Rockies and to have reached the Pacific Coast.

Now what is there to be said regarding the wisdom or unwisdom of the introduction of this little weaver bird into our country? Some years ago the attitude of the expert ornithologists of our United States Department of Agriculture was revealed by a sizable bulletin which was published under the title, "The English Sparrow as a Pest." This publication not only contained a summary of the food habits of the bird but also gave various instructions in regard to methods of control, even including poisoning. This last raised a storm of protests in various circles. These protests did not necessarily mean that the sparrow had hosts of friends but doubtless was due to the fear that other birds in addition to the sparrows would become victims of the poisoning operators. Many hundreds of times I have been asked the question, "What can I do to keep the sparrows away from my feeding station"? And "how can I prevent the English sparrows from taking possession of my martin house"? And I am compelled to reply that so far as is known nothing can be done. Now and then it helps a bit to either take your martin house down in the fall, not erecting it until the martins arrive in the spring, or else stop the openings at the end of the nesting season not reopening them until the advance scouts of the martins arrive.

Despite the general dislike of this little alien it has been proved to be of considerable value as a destroyer of many harmful insects such as cut worms and cabbage worms.

The nest is of almost any available material and is placed in every kind of situation. The eggs are 4 to 7, soiled white speckled with brown.

39

EASTERN SONG SPARROW
(Melospiza melodia melodia)

The song sparrow is found in its twenty-six described forms, all the way from Alaska to Florida and it is resident over much of its range, even braving snow and ice of winter. The eastern form, or type species, which will average six inches in length, is a resident during winters throughout a greater portion of the state. It is a hardy, resourceful bird and I have heard it, now and then uttering its silvery flute-like notes during a belated cold spell in late February or early March when the temperature was hovering around zero and there was a foot of snow on the ground. Heard thus on the bright and still winter morning its song breathed forth a note of optimism and good cheer and was a prophecy of better days to come when the pussy willows would be out and the hylas would be piping in the ponds.

The song sparrow adapts its mode of life to a great variety of habitats. This characteristic no doubt is responsible, during the long course of the centuries, for the differences, some of them slight to be sure, which have come about in various parts of the song sparrow's range. Thus in the moist climate of the Aleutian Islands with heavy rainfall, we have an exceedingly dark form, while in the arid and semi-desert region of our Southwest we find the desert song sparrow, which is exceedingly pale.

Not only is the song sparrow a hardy and adaptable bird but its behavior and home life have for years been the subject of intensive study and observation by ornithologists. Mrs. Margaret Morse Nice, a Fellow of the American Ornithologists' Union, has achieved much distinction by reason of her long continued studies of this sparrow and her published observations have been widely read by ornithologists and nature students.

The song sparrow, as is the case with many of our other song birds, is often imposed upon by the parasitic cowbird who deposits her eggs in the nests of other birds as a slip-shod and lazy way of avoiding the task of nest-building and later parental duties.

During the month of May one year I discovered a cowbird's egg in the nest of a song sparrow in the hedge which borders our garden. Having, for the time being, forgotten the incident, I was somewhat surprised two or three weeks later when I saw a lusty young cowbird on the lawn, not far from the nest where he had been hatched, being fed by one of his foster parents.

The song sparrow is one of our most useful birds for during the summer months its food consists almost entirely of insects. One time, for some minutes, I stood like a statue by a row of cabbages in our garden during which time a song sparrow examined each head, feeding upon the worms it found thereon.

The nest is either on the ground or in low bushes, of leaves, grass or strips of bark and lined with hair. The eggs are 4 to 5, light greenish or bluish marked with brown.

STARLING
(*Sturnella vulgaris vulgaris*)

One night in early winter, some years ago, as I walked down Pennsylvania Avenue in Washington, D. C., I was suddenly made aware of a continuous and multitudinous twittering which came from the buildings on both sides of the avenue and filled the air, even above the noise of the traffic on the street. Of course I was at once aware of the source of these raspings and twitterings which came from this great multitude of voices, for on many occasions I had watched the great flights of starlings as on winter evenings, they came in countless numbers to roost on Grant's Tomb on Riverside Drive in New York City and also on the Metropolitan Museum of Art.

In 1890 eighty starlings which had been brought over from Europe were released in Central Park, New York City. The next year forty more were brought over and released. It now appears that this little feathered alien is destined, one day, to occupy the entire continent. It is resourceful and aggressive and adapts its way of life to all sorts of conditions. Its first appearance in Indiana was in 1927 when a pair was found nesting between Anderson and Pendleton by Sidney Esten. Today one might hazard a guess that the starling population of our state exceeds that of any of our native resident birds. The introduction of the starling brings up the controversial question of the wisdom of introducing any alien species of bird or mammal into another continent.

Now and then such practice has been followed by a whole series of disastrous results many examples of which might be given. One was the introduction of the carnivorous mongoose from India to the Island of Jamaica with the thought that this cobra killing creature would help to rid the island of snakes. This was accomplished to some extent and then the mongoose became the single greatest destructive agent to the birdlife of the island. So whatever opinion we may hold as to the wisdom or unwisdom regarding the bringing of the starling to our shores, this bird is evidently here to stay and we had just as well make the best of it. Now for a moment let us examine its status. The two most common habits that have brought it into disrepute are first that of congregating in vast numbers in our cities and towns, during the winter months, and aside from the noise, the smearing of the building with their excrement. The other objection is their well known tendency to drive away our native hole-nesting and box-nesting birds by preempting their nesting places. Starlings are of considerable value as destroyers of harmful insects such as the Japanese beetle and many others. This little alien, which will average 8½ inches in length, now ranges from Canada and New England south to Florida and Texas westward to the states east of the Rockies.

The nest is in a hole in a tree or any kind of cavity or in bird boxes. The eggs are 4 to 6 and pale bluish.

RED-EYED TOWHEE
(Pipilo erythropthalmus erythropthalmus)

If you are a lover of the woodland ways and are familiar with the birds that commonly make their homes therein you surely cannot have failed to make the acquaintance of the towhee. He is a sturdy and conspicuous member of the finch or sparrow family and will average 8 inches in length. The male is really a bird of striking appearance with his coat of black, white and chestnut. The white markings on the tail are also a conspicuous field-mark as you surprise him in the thicket's shade and he takes wing to some nearby copse. The female is more soberly attired for she is a rich warm brown where the male is black. Our bird also is sometimes called ground robin, chewink and joree, the last two names having been given him on account of their resemblance to his notes. These have a ringing far-carrying quality about them and there are few dense woodlands or thickets within the state where they may

not be heard during the months of May and June. The towhee is for the most part a ground feeding bird and often as you quietly stroll through some dense wood, even on a winter day, you will hear his vigorous scratchings as he scatters the fallen leaves in his search of food. Now and then in case you surprise him he will leave his usual haunts and from some more elevated perch proceed to inspect you. This handsome bird is associated in my memory with early years on the farm near Monrovia when I roamed the woods of the hill country as a carefree lad whose chief purpose in life, at that time, was to learn to know the birds and other creatures of the woodland. In such haunts the two most conspicuous birds in the woods of May and June were the towhee and the yellow-breasted chat. Their notes had a far-carrying quality about them that rendered them audible for a considerable distance. Ernest Thompson Seton has interpreted the song of the towhee thus—*"drink your tee,* and *"chuck-burr, pill-a-will-a-will-a."*

The late Dr. Frank M. Chapman has given an excellent description of our bird as follows—"There is a vigorousness about the towhee's notes and actions which suggests both a bustling, energetic disposition and a good constitution. He entirely dominates the thicket or bushy under-growth in which he makes his home. The dead leaves fly before his attack; his white-tipped tail feathers flash in the gloom of his haunts. He greets all passers with a brisk inquiring *chewink, towhee,* and if you pause to reply, with a *fluff-fluff* of his short, rounded wings he flies to a nearby limb to better inspect you. It is only when singing that the towhee is really at rest."

The towhee is a common resident in the counties of southern Indiana. It is a useful bird as a destroyer of insects, most of which are injurious. The nest is on or near the ground, of dead leaves and grasses. The eggs are 4 to 5, white speckled with rufous.

TUFTED TITMOUSE
(Baeolophus bicolor)

The tufted titmouse is an outstanding member of the chickadee family which, has the distinction of possessing a conspicuous crest. It is sometimes called Tom tit, also sugar bird and Peter bird. The last mentioned name is derived from its oft repeated notes of *peter, peter, peter* or *peto, peto, peto.* It also has other notes. My first acquaintance with the titmouse was made as a farm lad when these birds were always common in the maple woods of late winter and early spring as preparations were being made to tap the trees to make the sap into both syrup and sugar. At such times the titmouse was by far the commonest bird and inasmuch as it was associated with the sugar making, it in many localities, was known by the above mentioned name. In favorite or congenial haunts one is often surprised at the number of titmice to be observed within a comparatively restricted area. One day in early April sometime ago the County Superintendent of Schools of Orange county, Mr. Orville Cornwell, accompanied me to the Cox-woods which is some two miles southeast of Paoli. While inspecting the walnut trees we also had our eyes and ears open for the birds that were inhabiting this tract of primeval woodland. Within the space of an hour eighteen tufted titmice were counted. The woods was ringing with their cheery far-carrying notes.

The titmouse is found not only in woodlands but it also comes about towns and villages. On one occasion I counted sixteen of these birds in the trees along the streets of a county seat town in the southern part of the state.

The titmouse is just as much of an acrobat as its cousin the chickadee and it often may be seen swinging from a pendent twig or hanging head downward from your feeder. Mention was just made of the numbers of titmice observed in spring time. Many of us also associate this bird with the crisp and frosty days of autumn when the nuts are falling and a hint of approaching winter is in the air. At such times titmice are often found in the company of jays, chickadees and nuthatches as they noisily go trooping through the woods.

The titmouse is regarded as a bird of considerable value from the point of view of its food habits. Government experts have found that more than half of its food consists of animal matter. Caterpillars and wasps make up more than 50 per cent of the food. It also feeds upon wild berries, seeds and acorns.

The tufted titmouse ranges throughout Eastern United States north to the southern part of New England and to southern Michigan and west to Nebraska and central Texas. It is an abundant resident in southern Indiana, rare farther north. It is from 5¾ to 6½ inches long. The nest is in a hole formed by a woodpecker or natural cavity in a stump or tree. The eggs are 5 to 6, creamy white and speckled with chestnut and sometimes lilac.

43

BLACK VULTURE
(Coragyps atratus atratus)

One day in late November, some years ago, I was tramping over an upland field on the home farm in Morgan county. A biting wind from the north was driving before it the first snow storm of the season. It was prophetic of the winter days that were just around the corner. I chanced to look upward and to my great surprise I saw three black vulture-like birds which were soaring and moving before the smiting blast, s o m e distance overhead. I watched them closely for a few minutes and soon discovered that the birds were not the more common turkey vulture but were black vultures or carrion crows whose principal range is farther south.

The black vulture is from 23 to 27 inches in length and has a wing-spread of from 54 to 49 inches. Any close observer may easily distinguish between these two vultures, even though they may be soaring at great heights, or are just small objects in the sky beyond. If not too far distant the under surface of the wing-tips of the black vulture appear a plain silvery gray. Moreover, the flight habit of the black vulture differs from its relative in that in the midst of its graceful soaring it resorts to frequent intervals of a succession of rapid wing-beats. This habit is all that is needed for quick and accurate identification. At quite close range it easily may be noted that the upper part of the naked neck and head are black whereas these parts in the case of the turkey vulture are bright red.

Vultures, wherever found throughout the world, are among nature's most persistent scavengers and as such perform a useful service, especially in those sections where sanitation is neglected and the disposal of the carcasses of dead animals is seldom or never practiced. It is no uncommon sight in some of the cities and towns of the South to see vultures of both species foraging about the streets for offal which has not been removed, or even sitting on the housetops dozing away the hours, perhaps in eager expectancy of some unforseen event which will provide them with a ready meal. The late Dr. T. Gilbert Pearson has related this incident: "One day a lady of my acquaintance, while sitting alone in her room, was much startled when a beef-bone fell down the chimney and rolled out on the hearth. Going outside she discovered a turkey buzzard peering down the chimney in quest of his prize." Vultures have been accused of disseminating hog cholera and other diseases of farm animals and this long has been a controversial question. It also, has been a moot question among ornithologists as to whether vultures discover their prey, which consists mostly of carrion, by sight or smell.

The nest is on the ground, under a log or bush, or in a hollow tree. The eggs are usually 2, grayish-green, marked with chocolate and reddish brown.

44

TURKEY VULTURE
(Cathartes aura septentrionalis)

The turkey vulture, generally called turkey buzzard, is a familiar bird to most persons within the wide extent of its range. No doubt its habit of feeding almost exclusively upon carrion has caused it to be regarded with something of a feeling of repulsion and disgust. However, it is one of nature's great inventions when considered from the viewpoint of its mastery of the air. Few will deny that the grace and beauty of its flight, in a measure, serve to redeem it and perhaps cause us for a moment to forget its lowly and disgusting feeding habits. Few summer skies within our state are without one'or more of these great birds, with their six feet spread of wings, silhouetted against the blue. Seen thus we are forced, at least into momentary admiration of a creature that long before the advent of the airplane had succeeded in a complete and perfect mastery of the air. As I write this there flashes across my vision a picture I shall never forget. It is a vision of turkey vultures soaring far aloft in the blue vault of heaven in company with stately wood ibises and water-turkeys. This was in the region of the great marshes of the upper St. John's River in Florida. During the heat of noonday I often watched them from my camp in the palm grove. Hour after hour these birds, caught and floating in the upper air currents, soared on motionless wings. It was both a beautiful performance and a perfect example of complete mastery of the air. However, of each of the soaring birds, I felt that the vulture was supreme as a perfect example of the ability of a bird to make itself at home in the upper air. Bradford Torrey once remarked that, "One might almost be willing to be a buzzard to fly like that". When, however, our gracefully soaring bird descends to mother earth it loses nearly, if not all, of the ethereal qualities we ascribe to it when in the sky for it becomes a lowly scavenger feasting gluttonously upon the carcasses of the dead animals it finds on farms or those killed by motor cars on our highways. It is no uncommon sight along the highways of the south, where there are no fences, to see both turkey and black vultures feasting upon the carcasses of cattle which have been killed by motor cars. Now and then the motorist comes to grief as in the case of a man who informed me that just as his speeding car passed a number of the birds that were gathered about a dead cow one of the birds arose immediately in front of his car and was killed as it smashed the wind-shield.

This vulture, which is from 26 to 32 inches long, ranges over most of the continent from southern Canada south to Mexico, Central and South America. It is resident most years in southern Indiana.

The nest is generally on the ground in hollow trees, logs, stumps, or occasionally in dense thickets. The eggs are usually 2, dull white with chocolate spots and blotches.

45

NORTHERN DOWNY WOODPECKER
(Dryobates pubescens villosus)

This active and energetic little member of the woodpecker family is the smallest of the group to which it belongs. It will average 6¾ inches in length which is about 3 inches shorter than its larger relative, the hairy woodpecker. With the exception of the matter of size and the black bars on the white outer tail feathers, it is almost an exact duplicate of the hairy. The males of both the downy and the hairy have red on the back of the head. This is entirely lacking in the females.

Probably most persons when they see any member of the woodpecker family, which is more or less barred or marked with black and white, immediately call it a "sapsucker." It should be remembered, however, that there is only one true sapsucker in eastern United States and this is the yellow-bellied. This is the bird which girdles our fruit and other trees with its countless borings which often damages the trees

The downy woodpecker is a general favorite with that large and ever-increasing number of persons who derive pleasure from feeding birds in the winter. It comes fearlessly and confidingly about the feeder even to one placed on the window sill. Its favorite food is suet and when it finds that a supply is always provided it comes almost every hour of the day to dine at this enticing board. When the thermometer is hovering around zero and the ground covered with a heavy snowfall and the trees all sheeted with ice, it is a source of much satisfaction to have not only the downy but other birds as well as regular visitors.

I once was guest in a home of bird-loving people. A window feeder, the top and inner side of which was made of glass, was installed on the window sill under the partially raised window. So my friends had birds literally coming and feeding in their living room. To this feeder the downy came as well as chickadees, juncoes and other birds.

Probably none of our native wild birds is more resourceful and hardy. Thousands of them brave the snow and ice and storms of winter throughout the whole of their more northerly range. Under such natural conditions they do a lot of scouting for food on their own account. A quiet walk in the winter woods will reveal them tapping for grubs and beetles and other insects which are concealed under bark or in the dead wood.

At all seasons of the year the downy woodpecker is a tireless searcher for many kinds of insects, tapping the cocoons of the coddling moth, eating the eggs of the canker worm and feeding upon weevils, ants and plant lice.

The northern downy woodpecker is resident throughout eastern United States south to eastern Nebraska, Kansas, Tennessee and Virginia. Two subspecies are recognized, southern downy woodpecker and Nelson's downy woodpecker. The nest is in a hole in a tree. From 4 to 6 white eggs are laid.

EASTERN HAIRY WOODPECKER
(Dryobates villosus villosus)

The hairy woodpecker is a conspicuous and interesting member of the family to which it belongs. In its several described varying forms it is distributed over most of the wooded areas of the United States and Canada. Our Indiana representative of the species which is from eight and one-half to nine inches in length, is found in Eastern United States from New England north to Nova Scotia, Quebec; west to Manitoba and Montana; south to Eastern Texas, South Atlantic and Gulf States. It is resident throughout most of its range.

In appearance the hairy closely resembles its near relative the downy woodpecker. In fact they look almost as much alike as two peas in a pod, only the downy will average some three inches shorter than its larger relative and in addition has black markings on the outer white tail feathers which in the case of the hairy are pure white.

The hairy woodpecker in comparison with its near relatives is a rather shy and suspicious bird and usually does not come so fearlessly and confidently to the feeding station. However, on many a bitterly cold winter day when the ground is blanketed with snow and the trees sheeted in ice, under stress of such conditions, he will come regularly to your feeding station. Often from our kitchen window we have watched him as he eagerly feeds upon the suet which we regularly keep in the feeder which is fastened on a big limb of the apple tree some ten feet from the window. At such times Mr. Hairy is the sole proprietor of the feeder, and all the other birds, including titmice, chicadees and starlings, depart in haste and nervously await a more favorable time for their return.

The call note of this woodpecker is rather loud and harsh and can be heard for a considerable distance. It is difficult to describe or imitate, but may be called a high pitched rather metallic *chink* or *click*. This may be heard for some distance. Sometimes it utters a succession of notes which somewhat resemble the flicker's *wick-a-wick-a-wick*. All its notes are much louder than those of the downy which they closely resemble.

This woodpecker, like most other members of the family is of considerable economic importance to farmers and fruit growers as well as being a patroller of our forests, helping to conserve them by reason of the great numbers of grubs, larvae and other insects which it destroys. It eats many injurious caterpillars, weevils, ants, plant lice and scale insects, as well as locusts, grasshoppers, spiders, and millipeds. It is also known for its ability to help rid a potato field of potato bugs.

The nest is usually in woodlands, rarely in orchards or near homes. The excavation, made usually in a dead tree, is about two inches in diameter and is from five to sixty feet up. It is from eight to sixteen inches deep and gourd shaped. The eggs are 3 to 5 and glossy white.

NORTHERN PILEATED WOODPECKER
(Ceophloeus pileatus abieticola)

Now and then your genuine bird enthusiast experiences a special thrill when he makes a discovery which he had least expected. Such was my good fortune not long ago. It was on the afternoon of a fine day in late October. Motoring along highway No. 37, some miles south of English, I suddenly saw two large black looking birds, the size of crows, which were flying across the highway only to disappear in the heavily forested hillside beyond. As the birds went galloping across the sky their un-_____ory flight at once identified them as pileated woodpeckers or, as this great bird is often locally called, woodhen and logcock. So I say it gave me much pleasure and satisfaction to note that this magnificent member of the woodpecker family had not yet been extirpated from our state and gone the way of some others of our rare, vanishing and extinct birds.

This woodpecker is from 16½ to 19½ inches in length, with a wingspread of from 28 to 30 inches. The stout beak is from 2 to 2½ inches long.

The flaming, poppy-red crest of the male gives a touch of distinction to this really handsome member of the family, while the female is similarly adorned with only about half of the terminal part of the crest being red.

In the early pioneer days, in the heavily wooded area of the state, the pileated woodpecker was no doubt a fairly common resident. Then came the clearing of the forests to make way for a diversified agriculture. This naturally brought about a disappearance of many of the larger and more conspicuous wild birds and mammals. Today, however, thanks to our fine system of State Parks and State Forests, combined with protective laws and the general change in public sentiment, the pileated woodpecker is undoubtedly making a comeback, and it is believed that a further increase will occur as time goes on. As recently as May 1947 a nest was found by members of the Indiana Audubon Society in McCormick's Creek State Park. There is not space here to list all the records that have recently come to me of the presence of this woodpecker in various parts of the southern half of the state. High school science teachers in Aurora, Cannelton and Sullivan have given me recent records as have some of the students. The ordinary call note of the pileated woodpecker is a loud and far carrying cack, cack, cack.

It may be gathered from the size of this woodpecker that it strikes powerful blows with its great beak, not only when it is searching for grubs and beetles in the dead wood of trees but also when it is busy excavating a cavity for its nest. Crawling to within fifteen or twenty feet from where one of these birds was busily engaged in making an excavation, I saw chips falling away in all directions.

The food consists largely of grubs, beetles and ants. In its four recognized forms it ranges over most of the United States and Canada. The nest is an excavation in a tree, usually well up. The eggs are 3 to 5 and glossy white.

48

RED-BELLIED WOODPECKER
(*Centurus carolinus*)

Most bird students will agree that of all our native woodpeckers none is handsomer or of more striking appearance than the red-bellied, or as he is sometimes called the Carolina and also the zebra woodpecker. Here is a bird whose color pattern is indeed striking and he is calculated to attract the attention of even the most indifferent and casual observer. The whole top and back of the head of the male is scarlet or bright poppy red. The female, however, differs from the male in that the red of the crown is lacking, occurring only on the nape and back of the head. Both sexes have the back conspicuously barred with black and white which has given the bird the name of zebra woodpecker.

To some this name of red-bellied woodpecker seems somewhat inappropriate inasmuch as the ordinary observer or even the trained bird student seldom gets a chance to note the faint wash or tinge of reddish on the bird's belly. The beautiful and conspicuous markings of this woodpecker, which averages 9½ inches in length have no doubt, at least in former years, made it a favorite target for thoughtless and irresponsible persons who have deemed it a good mark at which to shoot. It is nowhere as common as its relatives, the hairy and the downy woodpeckers, yet it occurs regularly throughout the southern and central portions of the state.

The call note of this woodpecker has always given me a special thrill whenever or wherever I have heard it, whether in the early spring in the beech wood of Indiana parks and farms or in the piney woods or hammocks of Florida. It is a vigorous, loud and far-carrying *tchurr, tchurr* or *chow, chow*.

The red-belly differs from the downy, in that it lacks the confiding and fearless disposition of its smaller relative. It usually appears rather shy and suspicious although it will often come to your feeding station, especially under stress of winter weather when snow and sleet cover the ground. During recent winters both male and female red-bellies have been fairly regular visitors to our suet container at Mooresville. No matter how frequent these visits may be they are always the occasion of a little excitement as we watch one of these truly handsome birds greedily feeding on the suet which has been provided for them as well as for other birds. While in the act of feeding the red-belly seems a bit nervous and ill at ease as though expecting some enemy to appear at almost any moment. So after swallowing great chunks of the suet he makes off for some secluded retreat.

This woodpecker, which ranges throughout eastern United States south to Florida and the Gulf Coast, is of considerable value as a destroyer of insects of which its food largely consists, such as beetles, roaches and caterpillars.

The nest is in holes in trees, stumps or posts. The eggs are 3 to 5 and dull white.

RED-HEADED WOODPECKER
(Melanerpes erythrocephalus)

It is generally agreed that the red-headed woodpecker is by far the most strikingly-attired of any of the family to which he belongs. With his tri-colored coat of bright crimson, bluish-black and white, he is calculated to attract the attention of even the most casual observer. The sexes are alike in color. The young in juvenal plumage have the red parts replaced with brownish-gray, or grayish-brown. The red-head is not only distinguished by his striking appearance but also by his ability to adapt his way of life to all sorts of conditions.

One bright and crisp October day not long ago I strolled leisurely over our farm in Morgan county. Here is a beechwood which from my earliest boyhood days has been a favorite haunt and home of the red-head, for beechnuts are his best-loved food. So on this recent visit I was immediately greeted with their noisy notes of *tchur, tchur,* from various parts of the wood. I soon discovered the red-heads had very important business on hand, for the beeches were loaded with

nuts and the birds were busily engaged in hoarding them for winter use. For this purpose many a crack and cranny or hole in the top of some decayed beech is utilized. It would be difficult to even hazard a guess as to the quantities of nuts so stored. In addition to beechnuts, acorns are stored in a similar manner. During years when the crop of beechnuts fail the red-head usually disappears during the winter, for he is irregularly migratory and adapts his habits to prevailing conditions. I recall, one winter some years ago, when my work was with the home office of the National Audubon Society in New York City, having almost daily observed a pair of red-heads that were wintering along the palisades of the Hudson River. They seemed to be faring well on the acorns which had been stored for winter use. They are expert fly catchers and during the summer months large numbers of insects of various kinds are eaten.

Although the red-head appears, in some respects, to be the most versatile of any member of the family, he has one incorrigible habit which, since the coming of the automobile, almost has been the cause of his undoing. I allude to the habit of frequenting telephone poles along our highways, from which vantage point he lights on the pavement to pick up various dead insects and other bits of food. Unlike some other birds, for instance the mourning dove and the crow, which invariably make a quick getaway, the red-heads have been killed in great numbers, especially on the highways of Ohio, Indiana and Illinois. Some years ago as we motored from New York to Indianapolis we recorded more than 1,000 birds and mammals dead on the pavement. Among these the red-head outnumbered all the other birds observed. This woodpecker, which averages 9¾ inches in length, is found in the eastern United States, ranging from southern Canada and New England to the Gulf of Mexico. The nest is in a dead tree or stub. Four to 6 white eggs are laid.

CAROLINA WREN
(Thryotharous ludovicianus ludovicianus)

All members of the wren family are extremely active and energetic birds, from the little mite of a winter wren to the largest member of the family, the Carolina, which is about 5½ inches in length.

Although modestly attired, the Carolina wren, with the rich rufous upper parts, cream-buff and white breast and conspicuous white line over the eye, is well calculated to attract attention at all times. To these characteristics should be added its far-carrying melodious notes, which may be heard during most months of the year. Certain of his notes remind one of the song of the Maryland yellow-throat, while others again call to mind those of the cardinal. One hesitates to say to what great distance his rich and tuneful song may carry, but when all else is still he easily may be heard at the distance of a quarter of a mile. In addition to his song he has a series of rather petulant, scolding notes which he seems to emit with greater vehemence and more frequently, in case you invade his woodland home.

He is both a dweller in woodlands and thickets and in towns and villages. Some times he even builds his nest in the woodshed or some other outbuilding. The writer recalls a nest which was placed in the opening of a sack of wool which hung in the tool shed on a farm in Morgan county. During extremely severe spells of winter weather when snow and ice cover the ground, this wren, as well as other winter residents, often suffer great hardship, even perishing from hunger and cold. I recall a little tragedy that occurred under such conditions some time ago. Two mouse traps had been set in a garage and baited with bits of butter. Some time during the bitterly cold day a pair of these wrens found entrance into the garage. On entering the garage the next day the owner found the two dead wrens, which had been caught in the innocent-looking traps. Needless to say he was much distressed, and no more mouse traps were set in the garage.

The food of this wren, as well as that of other members of the family, during a greater portion of the year, consists mostly of insects, nearly all of which are injurious. And it is amazing the great numbers that are consumed, not only by the young in the nest but also by the adult birds.

This wren is rather widely distributed, as it is found from northeastern Mexico and the southern states east of the Great Plains, and north to Nebraska and southern Michigan, southern Ontario and Connecticut. It is resident throughout its range. There is some evidence to the effect that during recent years it has been extending its northward range in Indiana. The nest is built in a variety of places, such as brush piles, holes and crevices in trees or stumps or in bird houses. It is rather bulky, of weeds and grasses, wool, rags, etc. The eggs are 4 to 6, white speckled with brown.

51

FEATURES

The articles appearing on the next three pages were prepared by Alden H. Hadley, for the purpose of providing worth-while recreational projects for youngsters, as well as adults, who are interested in the study of our feathered friends.

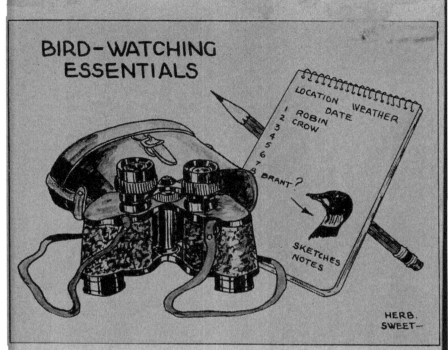

BIRD-WATCHING ESSENTIALS

LOCATION WEATHER DATE
1 ROBIN
2 CROW
3
4
5
6
7
8 BRANT?
SKETCHES NOTES

HERB. SWEET—

During recent years the term-bird watcher has come into common use as a substitute for the older and perhaps overworked one of bird-lover, which perhaps smacks a bit of undue sentimentalism. I believe we have borrowed the newer and more appropriate name from our English cousins who are inclined to be a little more matter of fact.

At any rate, a bird-watcher is a person who presumably has a special interest in birds and who may be any individual from a learned fellow of the American Ornithologists' Union to a member of an Audubon Junior Club or a Boy or Girl Scout.

Now the essentials of a good bird-watcher are, first of all (which is of course taken for granted) an abounding never flagging interest, then good, keen eyes and ears, which enable the watcher to discriminate between sounds, color and form. A field glass of some reliable make is absolutely essential. Then your really seriously minded bird-watcher will record his or her daily observations in a note book which is carried for that purpose.

The amateur bird student will doubtless make more headway in bird identification during the winter months at which time the birds are more easily observed and the confusion due to the greater number of species is not present.

Well do I recall, as a lad in the early teens, the bewilderment and confusion which I experienced during the height of the spring migration when the great hosts of feathered migrants came streaming northward from summer climes. The confusion was even greater during the fall migration on account of the many and varied plumage changes in the birds.

After fair progress has been made in the field of bird identification the serious minded student will do well to turn his attention to bird behavior for here is a field of unlimited possibilities. How little, after all, is known concerning the complete life history of even our most common birds!

53

FEEDING STATIONS

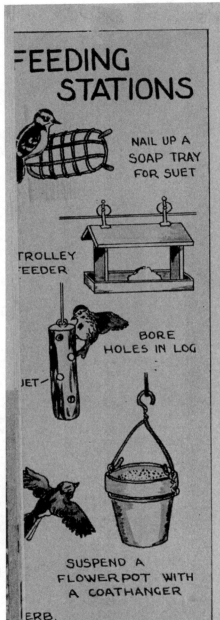

NAIL UP A
SOAP TRAY
FOR SUET

TROLLEY
FEEDER

BORE
HOLES IN LOG

JET—

SUSPEND A
FLOWERPOT WITH
A COATHANGER

ERB.
NEET—

Some years ago the National Audubon Society received a letter from a member who deplored the efforts of the Society in encouraging its members to attract birds about their homes by winter feeding and also by the provision of nesting boxes. Her objection was that such efforts tempted cats to catch the birds! Now if you will just glance at the cuts on this page you will note that three of the feeders are suspended and are therefore cat-proof. The trolley feeder is ideal for the feeding of various grains, bread-crumbs, etc. Not only does the roof protect the food from the weather, but with a string attached it easily may be drawn into your porch or window in order to replenish the supply of food.

One must not fail to keep the suet-feeder filled, for this is often a life saver for the birds under stress of snow and sleet and bitter weather.

Cold alone seldom kills birds, but failure to find food lowers their vitality any they often die of starvation when the food supply is not to be had. I recall having been caught in a midwinter blizzard a few years ago in the northern portion of the state. More than a foot of snow had fallen. This was followed by a sleet storm and the fields, woodlands and countryside were sheeted in ice.

Under such conditions many birds, both game and non-game, inevitably die of starvation. In case one buys his feeding devices from a dealer in such things there is usually a wide choice both as to cost and kinds. However, one need not go to much expense. A strip of woven galvanized wire with the mesh about ½ inch in diameter will serve admirably as a suet feeder.

It is not necessary to buy expensive mixtures of food for birds. For the chickadees, nuthatches, titmice, brown creepers, and hairy and downy woodpeckers, the suet feeder is a favorite haunt, while cardinals and other seed-eating birds find the trolley and window shelf feeder more to their liking. Remember, provision means friendliness.

SIMPLE BIRD HOUSES

— USE SLAB WOOD WITH BARK

TACK UP AN OLD HAT —

TIE UP A GOURD

WIRE A FLOWER POT TO A BOARD·

HANG IN A SHADY SPOT

° RELIEVE THE HOUSING SHORTAGE ·

HERB. SWEET—

Probably the simplest definition ever given of a bird-sanctuary was that of the small boy who said: "You put out some suet in the winter and some nesting boxes in the spring and then you 'shoo' the cat away."

This is as much as to say that provision means friendliness and it is surprising what may be accomplished in attracting birds about the dooryard by the provision of suitable types of nesting boxes. It must be remembered, however, that some birds are strict observers of what ornithologists call "territorialism" and are very jealous of the intrusion of other members of their species. Starlings and English sparrows are an ever-present problem and menace and there is little or nothing that can be done about it for they easily find access into a martin or blue bird house. Sometimes it helps if these boxes are taken down in the fall and not erected until the first martin appears in the spring.

Here is one important "don't." Never construct a bird house of tin or metal.

Some years ago a maker of bird houses endeavored to interest the National Audubon Society in approving and in handling his product which was constructed of heavy tin. Needless to say he was greeted with a very decided "thumbs down" on his proposition and, it is hoped, he departed a much wiser man in this particular field. Of course it is possible that a bird house made of tin or crockery, if hung in a very shady spot, might be safely occupied by the birds.

Among the many thousands of people who find pleasure and satisfaction in putting out nesting boxes for birds, it is probable that homes for purple martins and house wrens find first place in their choice. A martin house should always be erected in an open space, not close to trees or buildings, and should have a cat guard on the pole. Wren houses may be placed almost anywhere.

5- 1084
Qu

VISIT INDIANA STATE PARKS

Your own Indiana state parks provide natural cover and habitat areas for many of the birds described in this booklet. Visit any of the fifteen state parks and study the birds and wildlife you find there. Use this publication as a guide to the birds you discover and identify them by referring to the photos and copy herein. For location of Indiana's state parks see the chart below.

Bass Lake Beach	Winamac
Brown County State Park	Nashville
Clifty Falls State Park	Madison
Indiana Dunes State Park	Gary
Lincoln State Park	Lincoln City
McCormick's Creek State Park	Spencer
Mounds State Park	Anderson
Pokagon State Park	Angola
Shades State Park	Waveland
Shakamak State Park	Jasonville
Spring Mill State Park	Mitchell
Tippecanoe River State Park	Winamac
Turkey Run State Park	Marshall
Versailles State Park	Versailles

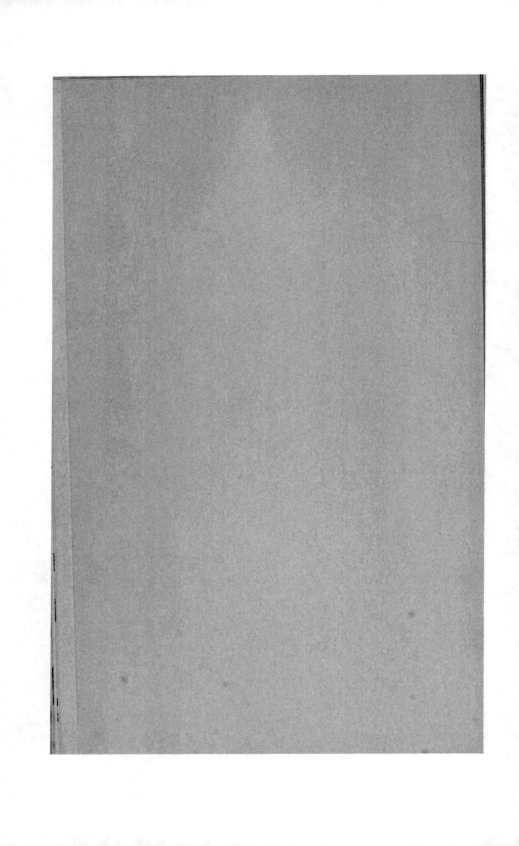

CPSIA information can be obtained
at www.ICGtesting.com
Printed in the USA
LVHW080447260323
742605LV00004B/50